KEEN, A. Myra and Eugene Coan. Marine molluscan genera of western North America; an illustrated key. 2nd ed. Stanford, 1974. 208p il tab bibl 73-80625. 8.75. ISBN 0-8047-0839-8

A completely revised edition of a successful reference work and handbook first published in 1963. The expanded coverage now permits, by means of beautifully illustrated keys, the identification of all shelled molluscan genera of the west coast of North America. Also, the glossary has been enlarged, and references to the recent literature are included. The work is recommended as essential for all west coast college libraries and for institutions in other parts of the country where work in marine biology is being carried on.

MARINE MOLLUSCAN GENERA OF WESTERN NORTH AMERICA

A. Myra Keen is Professor of Paleontology and Curator of Malacology Emeritus at Stanford University. Eugene Coan is a Research Associate of the Los Angeles County Museum and is on the staff of the Sierra Club.

MARINE MOLLUSCAN GENERA OF WESTERN NORTH AMERICA

An Illustrated Key

SECOND EDITION

A. MYRA KEEN & EUGENE COAN

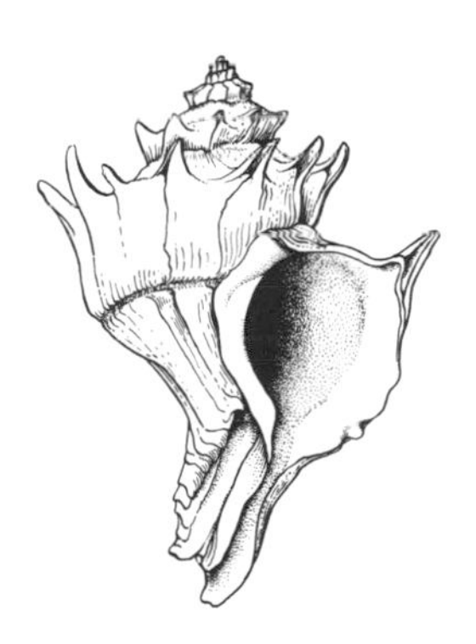

STANFORD UNIVERSITY PRESS
STANFORD, CALIFORNIA
1974

Some of the material in this book, in original or revised form, is from A. Myra Keen and Don L. Frizzell, *Illustrated Key to West North American Pelecypod Genera* (1939; rev. ed. 1953), and from A. Myra Keen and John C. Pearson, *Illustrated Key to West North American Gastropod Genera* (1952).

Stanford University Press
Stanford, California

Printed in the United States of America
Original edition 1963 Second Edition 1974
ISBN 0-8047-0839-8
LC 73-80625

ACKNOWLEDGMENTS

To the late Dr. Don L. Frizzell we owe the insight that molluscan taxa may be keyed out objectively at the generic level without regard to systematic affinities. He had demonstrated this in work with West American pelecypods (bivalves). Later, Mr. John Pearson was successful in extending the keys to include gastropod genera. We thank both of them for their willingness that their keys be even further expanded.

Several colleagues have read parts of the text for this edition as well as the earlier versions and have supplied us with data. Dr. James H. McLean has been particularly helpful with information on prosobranch gastropod taxa, making available to us lists and notes that result from the revisions he has made in preparing a modern manual on West American mollusks. Opisthobranch gastropods are attracting increased attention from students in recent years; we are indebted to several workers for advice on these groups, notably to Messrs. Don Cadien, James Lance, and Stephen Long. Mr. James Carlton, who has been preparing molluscan keys at the specific level for another work, not only pointed out a number of places in which our key choices needed clarification but also compared our generic lists against his to assure uniformity of interpretations. The key to the Polyplacophora benefited from the scrutiny and suggestions of Dr. S. Stillman Berry and Mr. Allyn G. Smith, both of whom also provided information on ecology and taxonomy and contributed to our summaries of the Cephalopoda.

Virtually all of the pelecypod illustrations are reprintings of figures drawn for the previous edition by Mr. Perfecto Mary, Senior Draftsman in the Department of Geology, Stanford University. He has also prepared a number of the gastropod figures, notably a new one of *Aclis (Graphis)* from a photograph by Dr. McLean. The major part of the gastropod figures were the work of Mr. Pearson, who, in order to assure accuracy, made ink-line drawings directly on photographs, later bleaching them to remove unneeded shading. The new figures of this edition are mainly photocopies from the literature. Mrs. John D'Aiuto assisted in bringing these into harmony with previous figures by skillful retouching and line emphasis.

For all of this help we are grateful, as we are also to all those persons whose contributions, though substantial, were less obvious. During the years since the first set of keys was published—in the late 1930s, when Elsie and Emery Chace took time to provide ecologic data from their extensive collecting experience—up to the present time, collectors have continued to supply us with information or have offered valuable criticisms. Any errors or oversights that remain in the work are surely much fewer because of such cooperation.

Responsibilities in preparing the sections of this book have been divided as follows:

Keen: Keys; Systematic Lists to Gastropoda, Pelecypoda, Polyplacophora, and Scaphopoda; Identification of Figures.

Coan: Ranges and Habitats; Systematic Lists to Cephalopoda and Aplacophora.

The Bibliography and the Glossary are joint efforts, and each coauthor has acted as critic of the other's work.

A.M.K.
E.C.

November 1973

CONTENTS

MARINE MOLLUSCAN GENERA OF WESTERN NORTH AMERICA

INTRODUCTION

Who has not, on first sight of a strange new animal, plant, or shell, said, "What is it?" or "What's its name?" For a time one may be content with a rough-and-ready common name like "thorny oyster" or "sundial shell," but sooner or later a need surfaces for finding out the corresponding scientific (zoological) name. Two distinct processes, identification and classification, are involved in this quest, and the two terms are often incorrectly used as if interchangeable. Identification is essentially the matching of an unknown specimen with a picture, or description, or physical specimen of a unit already named and established. Classification, by contrast, is the pigeonholing of that named unit within the larger framework of the known organic world. Identification may be attempted and more or less successfully accomplished by anyone with a good library or reference collection at hand. Classification, at least in its finer points, is the preserve of those trained in systematics, who are accustomed to taking into account much more than just the superficial appearance of a specimen.

Identification proceeds by stages. First, if the specimen is an animal, we establish whether it is vertebrate or invertebrate; and, if invertebrate, to what phylum it belongs (classifiers now recognize some 26 living phyla). Assuming that the phylum turns out to be Mollusca, we must then decide in which class it belongs (although many texts mention only five classes, modern workers recognize seven among the living mollusks and several others among the fossils). At this point most handbooks gloss over the next steps in the classification hierarchy, the order, family, and genus, to concentrate on listing as many units as possible at the lowest level, the species. Here the beginner can waste a great deal of time leafing through books to find likely pictures. The experienced worker, at the same time, may be unaware of this inadequacy in the literature, because practice in identification has already supplied a stock of mental images of shell forms, descriptive formulae, and names, against which, almost without thought, the new material is compared, to assign it to family or genus. The quick recognition, then, of generic (and hence familial) affinities marks off the expert from the novice. The present work has the aim of bridging the gap between the broad-scale popular handbooks and the fine-scale technical or local monographic works, to furnish the less experienced with something of the contents of the mental kit-bag of the specialist (and also, perhaps, to refresh the flagging memory of the seasoned worker). The book is therefore intended not as an introduction to the study of mollusks but as an intermediate-level guide for the identification of West North American material. And because most of the genera present on this coast occur also in other areas of the world, the book may be expected to have application beyond our chosen geographic limits. Nominally, the coverage remains the area from San Diego, California, north to the Arctic Circle, but actually it could well be said to include, southward, all of the Californian province, which extends along the outer coast of Baja California at least as far south as Isla Cedros.

Books and articles that are useful at the species level of identification are cited in the Bibliography of this volume. Some of the readily obtainable elementary and second-level guides that will help the beginner to a clearer idea of where West American Mollusca fit into the larger picture of marine invertebrates in general are listed below, with notes on their special features.

(a) Popular guides to the marine world in general:

R. W. Burnett, H. I. Fisher, and H. S. Zim. *Zoology, an Introduction to the Animal Kingdom.* New York: Golden Press (a Golden Nature Guide), 1958. 160 pp., 350 color figs. $1.00.

J. W. Hedgpeth. *Seashore Life of the San Francisco Bay Region and the Coast of Northern California.* Berkeley: University of California Press, 1962. 136 pp., 8 color pls., 88 text figs. $1.95.

Myrtle Johnson and H. J. Snook. *Seashore Animals of the Pacific Coast.* New York: Macmillan, 1927. 659 pp., 700 figs., 11 color pls. (reprinted 1967, Dover Books, $3.75).

H. S. Zim and Lester Ingle. *Seashores.* New York: Golden Press (a Golden Nature Guide), 1955. 153 pp., 475 color figs. $1.00.

(b) Technically advanced guides to the marine world or to invertebrates:

Libbie Hyman. *The Invertebrates.* New York: McGraw-Hill, 1940-67. Vol. I, Protozoa through Ctenophora. II. Platyhelminthes and Rhynchocoela. III. Acanthocephala, Aschelminthes, and Entoprocta. IV. Echinodermata. V. Smaller Coelomate Groups: Chaetognatha, Hemichordata, Pogonophora, Phoronida, Ectoprocta, Brachiopoda, Sipunculida. VI. Mollusca. [Comprehensive reviews of invertebrate morphology, with excellent bibliographies in each of the six volumes.]

S. F. Light, Ralph Smith, F. A. Pitelka, D. P. Abbott, *et al. Intertidal Invertebrates of the Central California Coast (S. F. Light's Laboratory and Field Text in Invertebrate Zoology)*. Berkeley: University of California Press, 1954. 446 pp., 138 figs. [A new edition of this standard reference work is in preparation.]

E. F. Ricketts and Jack Calvin. *Between Pacific Tides* (4th ed., revised by J. W. Hedgpeth). Stanford, Calif.: Stanford University Press, 1968. 614 pp., 8 color pls., 302 figs. $10.95.

(c) Popular-format, elementary-level guides to the Mollusca:

R. T. Abbott. *How to Know the American Marine Shells*. New York: New American Library (a Signet Key Book), 1961. 222 pp., 12 color pls., 159 figs. $1.25.

R. T. Abbott. *A Guide to Field Identification: Shells of North America*. New York: Golden Press, 1968. 280 pp., 155 color pls. $3.95.

R. T. Abbott. *Kingdom of the Seashell*. New York: Crown, 1972. 256 pp., 92 black-and-white pls., 178 color pls., 13 figs. $14.95.

R. T. Abbott and H. S. Zim. *Sea Shells of the World*. New York: Golden Press (a Golden Nature Guide), 1962. 160 pp., 790 color figs. $1.25.

P. A. Zahl. "The magic lure of shells," *National Geographic*, CXXXV, no. 3 (March 1969), pp. 386-429.

(d) A popular-format, advanced-level guide to the Mollusca:

J. E. Morton. *Molluscs: An Introduction to Their Form and Functions*. New York: Harper Torchbooks (Science Library), 1960. 232 pp., 23 text figs. $1.40.

(e) Treatises on Systematics:

Ernst Mayr. *Principles of Systematic Zoology*. New York: McGraw-Hill, 1969. 428 pp. $14.20. [An advanced textbook of taxonomic methodology.]

E. T. Schenk and J. H. McMasters. *Procedure in Taxonomy* (3d ed., revised and enlarged by A. M. Keen and S. W. Muller). Stanford, Calif.: Stanford University Press, 1956. 149 pp. [Out of print but available in libraries; its compact text still serves as an introduction to taxonomic methods, but the Appendix is now superseded.]

Keys and their development. At least as early as the 1790's authors were analyzing biologic groupings in charts or tables (for example, Cuvier's "Tableau élementaire," in 1797). These efforts clearly foreshadow our own modern keys and were mostly of a type that pursues, in an "either/or" set of choices, the "either" line to its end before taking up the "or" line. Customarily, such keys use letters of the alphabet, Roman numerals, or other symbols to mark off the corresponding sets of alternatives. They are flexible in that they can accommodate to the technically objectionable and grammatically incorrect "either/or/or" type of choice, but they can be confusing if they run to more than a single page. For a long key with many sets of choices, it is preferable to keep the pairs of "either/or" alternatives in a numbered sequence. That is the type of dichotomous key adopted here.

The first published key to any group of West North American molluscan genera seems to have been one by Mr. A. M. Strong, on microscopic gastropods, issued by the Conchological Club of Southern California in 1934.

The present keys derive from a student project undertaken by the late Dr. Donald L. Frizzell while an undergraduate at the University of Washington. He drafted a key to the pelecypod genera of Puget Sound, on the suggestion of a professor who felt that generic taxa were clearcut enough as concepts to be usable. The idea proved to have merit, and Frizzell later, when a graduate student at Stanford, sought collaboration for a key of broader geographic coverage—the pelecypod genera of the entire West Coast of North America. Trials of proposed drafts with other students soon showed that words alone were not enough, that the final choices needed to be confirmed by pictures. The illustrated version that resulted was published by Stanford University Press (Keen & Frizzell, 1939).

Users of this pelecypod key urged the preparation of a companion work on gastropod genera. Another Stanford student, John Pearson, undertook this formidable task and completed a rough draft and most of the illustrations before leaving for commercial employment. This, too, was published by Stanford University Press (Keen & Pearson, 1952).

After both keys had been reprinted and were needing revision, they were updated and reassembled into a single volume (the first edition of the present work, published in 1963), with keys to other groups (notably the chitons) added.

The present edition embraces further expansions. Although the keys themselves still are only to the shelled forms, the existence of a substantial number of non-shelled groups is recognized, not only in the list given under "Limits of Coverage" below but also in the Systematic Lists and in the section on Ranges and Habitats. We believe that in this new edition we have a complete or nearly complete listing of published generic records for the West Coast of North America, for all molluscan classes.

The Systematic Lists. As has been shown above, keys are useful devices for identification. They are, however, artificial in the sense that they merely point out visible differences between objects that at first glance may look very much alike. Because these differences are not necessarily a reflection of basic morphologic or phylogenetic characters, the groups that key out together may not be at all closely related. The sequence of genera in the keys, therefore, gives only the sketchiest of clues to true relationships. What is needed is a recapitulation in terms of modern biological classification; the Systematic Lists fill that need. The sequence of units in a sound classification should reflect (theoretically, at least) evolutionary advances from simple to complex; practically, most classifications fall short of this goal, but workers by and large are in agreement on general patterns. Thus, one expects to find, near the beginnings of lists, the taxodont bivalves and the round-mouthed nacreous-shelled gastropods, and, toward the ends, the wood-boring bivalves and the gastropods that have evolved highly modified fanglike radular teeth.

Genera in the systematic lists are grouped into families. For greater precision in indexing, the families are numbered here, sequentially by class.

Representatives of all seven living classes of the Phylum Mollusca occur in some part of the Eastern Pacific, but only four are treated in detail here: the Gastropoda, Pelecypoda (Bivalvia), Scaphopoda, and Polyplacophora (chitons). The Aplacophora (grouped along with the chitons, by some authors, as Amphineura) are mostly offshore in occurrence, and they lack any hard parts. The Monoplacophora have been recorded to date only in abyssal depths outside the geographic limits of this work, off the Hawaiian Islands and to the southward of Baja California. The Cephalopoda, also largely lacking hard parts, are currently being studied by several specialists, and we may anticipate considerable additions to the list of cephalopod taxa when their work is published. And again, the keys are to genera of shelled mollusks—the many non-shelled gastropods, for example, are listed in the Systematic Lists (chiefly under Opisthobranchia) but are not keyed out.

Ranges and Habitats. This section has been expanded to include data on the non-shelled mollusks. As before, the general distribution, the approximate number of included species, and the expectable habitat are given. Depth data for offshore forms now are in meters rather than fathoms. Superscript reference numbers after some of the names are explained under "Conventions Adopted" below. This tabulation of ecologic data should be especially useful to paleontologists and to those who are making environmental surveys.

Identification of Figures. Previous editions of the keys have not cited species names for any illustration. The line drawings for many of the genera are deliberately generalized or made composite and thus do not illustrate any one species. Often, however, a species may be selected as characteristic of a genus, or two species may be picked out that represent extremes of variation. These are now cited by name in a new section, "Identification of Figures," that precedes the Glossary. Reference citations for the sources of the figures are not given, but author and date for the species are shown in the conventional manner. Nearly all the figures are of Eastern Pacific specimens; exceptions are so noted.

Glossary. The glossary has been considerably broadened in this edition to include definitions of every technical term used in the keys. It includes also a number of common terms that systematists working with mollusks may use in a special sense.

Bibliography. Bibliographic references cited in earlier editions are augmented here not only by many more modern publications but also by citations of works useful in the identification of groups not previously canvassed, such as the Nudibranchia and the Aplacophora.

The only extensive checklist for this area (Dall, 1921) is now more than fifty years old; a somewhat revised and expanded version was published by Burch (1944-46). The publication of several molluscan volumes of the "Treatise on Invertebrate Paleontology" (Moore et al., 1960; 1969-71), and of the International Code of Zoological Nomenclature (1961; 1964), provides a background for more coherent taxonomic work than was possible hitherto, but there still remains a need for a modern review of West American mollusks at the species level.

Conventions Adopted. The use of italic type for generic and specific names, roman type for all other zoological names, parentheses around subgenera, and parentheses around the name of an author when a species has since been transferred to a different genus are obvious conformities to the rules of the ICZN. Some other usages, peculiar to this book, may not be of readily evident meaning:

*/ An asterisk indicates that a given genus in the keys has been keyed out twice (and is so indicated on both occurrences).

=/ An equality sign shows that a name is equivalent to a prior or a preferred name; the second name, in other words, is a synonym.

?/ A question mark standing ahead of an equality sign shows that the suggested synonymy is only tentative; standing after an equality sign but before the name of a taxon, a question mark shows that the systematic position or validity of the unit is in doubt.

" "/ Quotation marks indicate that a unit is cited as authors have used it in the literature, but that further research may well establish it as representing a misidentification.

Limits of Coverage. As old collections are studied and new specimens are obtained, the early records of genera must be reviewed. The genera that seem to have a dubious claim to authenticity or that are not truly endemic West North American forms may be grouped into several categories. These are signalized in the Systematic Lists and in the lists of Ranges and Habitats by the use of a superscript numeral, as follows:

[1]/ Migrants from adjacent provinces (mostly from the south), occurring intermittently and probably not adapted for permanent establishment. Included are these taxa, some of which are keyed out: *Anadara*, *Barbatia*, *s.s.*, *Lyropecten*, *Aligena*, *Raeta*, and *Gastrochaena*. The gastropod *Muricanthus* Swainson, 1840 (not keyed) has been collected but not formally reported, and doubtless there are others.

[2]/ Genera definitely occurring in adjacent areas but for which positive records for the West Coast have not yet been published. Units in this category are given in the Systematic Lists but not in the Keys: *Sarepta*, *Spinula*, *Cymatioa*, *Abra*, *Halicardia*, *Fossarus*, *Sigaluta*, *Tractolira*, *Crassispira* (*Crassispirella*).

[3]/ Doubtful early records, not confirmed in recent years. Included here are *Isognomon*, *Pteria*, "*Anisodonta*," *Fossarus*, and *Gibbula*, none of them cited in the Keys.

[4]/ Taxa valid as West Coast groups, but for which names currently in use are open to question. Examples are: "*Lepton*," "*Pseudopythina*," and "*Pleurotomella*," all in both the Keys and the Systematic Lists and cited in quotation marks.

[5]/ Imported groups, representative species of which have entered the area as by products of commercial activities and have become established in Western American waters. These include, among gastropods, *Cecina*, *Batillaria*, *Urosalpinx*, *Busycon* (*Busycotypus*), *Nassarius* (*Ilyanassa*), *Aeolidiella*, and possibly *Phytia*; and, among the pelecypods, *Ischadium* (*Geukensia*), *Crassostrea*, *Gemma*, *Tapes* (*Ruditapes*), *Petricola* (*Petricolaria*), *Theora*, *Laternula*, *Lyrodus*, and probably *Teredo*. Too late for inclusion in the Keys, a new record has been reported while this book was in press—*Sabia*, a limpet-shaped Western Pacific gastropod.

A sixth category might be made for those mollusks that are authentic but that lack any vestige of shell. This category comprises representatives from three of the West Coast classes (the sequence of the Systematic Lists is followed here):

Gastropoda
- Atlantacea (i.e., heteropods): family Pterotracheidae
- Parasita (all families)

Thecosomata: families Cymbuliidae and Desmopteridae
Cephalaspidea: *Runcina*
Anaspidea: *Phyllaplysia* (occasional specimens)
Gymnosomata (all families)
Notaspidea: *Pleurobranchaea*
Sacoglossa: Plakobranchacea (=Elysiacea, auctt.) (all families)
Acochlidioidea (all families)
Nudibranchia (all families)
Aplacophora (all families)
Cephalopoda: all except *Argonauta*

As a final category, we may cite in two groups those taxa that are entirely excluded from treatment herein. The first of these are nonmarine taxa, including those fresh-water residents like *Corbicula* Megerle von Mühlfeld, 1811 (a clam imported from the Orient) that might wash downstream to the coast, and brackish-water forms, such as *Tryonia* Stimpson, 1865, a small gastropod found in a few California lagoons. In a second group are erroneously reported records, such as *Anaplocamus* Dall, 1895, an East Coast fresh-water gastropod that was by some strange mischance described as occurring in Alaska; *Gibbula* Risso, 1826, a Mediterranean gastropod genus reported from California but not subsequently confirmed as native to the Eastern Pacific; and *Darina* Gray, 1853, a South American pelecypod mislabeled as from Puget Sound.

Problems of Taxonomy. To include in the Keys all named groups ranked at present as subgenera might be helpful to some students, but it would increase the size and complexity of the keys beyond reasonable limits and perhaps beyond maximum usefulness. A pragmatic approach has been taken here. Only the subgenera having morphologically distinctive characters that might cause problems are keyed separately. For example, a subgenus with strong radial ribs in a genus of otherwise smooth-surfaced forms could cause perplexities if a key choice were based on sculpture alone. Unless the genus is to be keyed out twice, such a subgeneric group ought therefore to be keyed out separately, its subordinate status indicated by the use of parentheses, both in the Keys and in the Systematic Lists. The opportunity has been taken in the Lists to cite many of the subgenera in current use for the West North American fauna, but an exhaustive census of all these seems beyond the scope of this work.

A second kind of problem arises with generic names inherited from early authors, names that in some cases may have been misapplied. Modern distributional evidence may show that a genus is otherwise restricted to some distant province. For example, Philip Carpenter, the first serious student of our fauna, described a small clam as a species of what he took to be *Lepton*. Now *Lepton* is predominantly a European group—small white clams that only superficially resemble Carpenter's type specimen (which unfortunately is an incomplete and somewhat broken shell). Lacking a suite of material that would establish clearly the expectable variation of the West Coast form, we retain the name Carpenter chose, but the use of quotation marks shows that the assignment is open to question. There are several such problem names in the Keys and Lists, the clarification of which will be challenging problems to future students.

In other groups, one encounters problems that are insoluble at present. Relationships may not yet be clear, or it may not yet be possible to distinguish with confidence between various subgroups. In some cases, we do not know whether to recognize many subgenera within a few broad genera or numerous genera within a family. In this book we have dealt with such problems largely by *ad hoc* judgments. Definitive solutions must await the efforts of future workers who, it is to be hoped, will have access to more and better material than has yet been available. Obviously, more than a little revisionary work remains to be done in the field of molluscan systematics.

KEYS TO THE GENERA

As in earlier editions of this work, the keys are dichotomous—i.e., they are based on successive dual alternatives. To help in backtracking, the number of the preceding alternative has been included in parentheses following the key-pair number at the beginning of each pair of choices. Thus, for example, in Gastropoda, choice 1 leads to choice 2 or to choice 43. If we turn then to these two alternatives, we see that the number in parentheses following both 2 and 43 shows that we have arrived at those points from choice 1.

The arrangement follows no set order of classification but is a purely artificial guide for separating groups of shells that show morphologic resemblances. After a tentative identification of genus is made, one should examine the illustration to make sure that the result is plausible. If not, backtracking may reveal some spot at which a wrong alternative might have been selected. If the identification, however, seems plausible, handbooks or other literature suggested in the Bibliography should be consulted.

Some species within a genus (or even specimens within a species) may lack a character that otherwise distinguishes the group. Genera known to contain such aberrant members are keyed out twice. Duplication of this sort is signalized by an asterisk.

The approximate magnification or reduction of the illustrations is shown by the conventional × symbol, rather than by the scales that were provided in earlier editions. As before, the stated sizes must be regarded as only a general guide, for there is much variation in the sizes of adults of any population.

KEY TO THE GASTROPODA

This key is not complete for all Western North American gastropod genera because there are a number (listed in the Introduction) that entirely lack any shell. A few genera have rudimentary internal shells that are not often seen; and one, *Phyllaplysia*, may have an internal shell in some individuals but not in others. Because this is a key to the shells, soft parts are mentioned only incidentally; however, a distinctive operculum or radula may be cited as a secondary characteristic feature.

Only mature specimens, reasonably well preserved and unbroken, are considered eligible for identification. No key can be infallible for beachworn or juvenile material. Size ranges for adults vary so much that the terms denoting size must not be defined or interpreted too rigorously. In general, "minute" shells are less than 5 mm in greatest dimension (usually height, i.e. length), "small" shells are 5 to 15 mm, "medium sized" 15 to 40 mm, and "large" over 40 mm. For shells of lenticular shape, in which height (or length) is less than width (or diameter), it is more convenient to refer to the larger dimension (i.e. diameter) when size comparisons are made.

No West Coast specimens or figures were available for a few genera, e.g., *Onchidiopsis*, and the outside source of these illustrations is so indicated in the "Identification of Figures" listings, pp. 165-70.

Some gastropod groups, notably the Turridae, are much in need of systematic work; studies of radulae, in particular, have demonstrated this need. At present, the distinctions made between many named generic and subgeneric groups on shell characters alone seem unclear. Therefore, pending thorough revisions, several of these taxa are keyed out in only a broad sense, with bracketed notes to indicate possible relationships. Deepwater forms pose special problems.

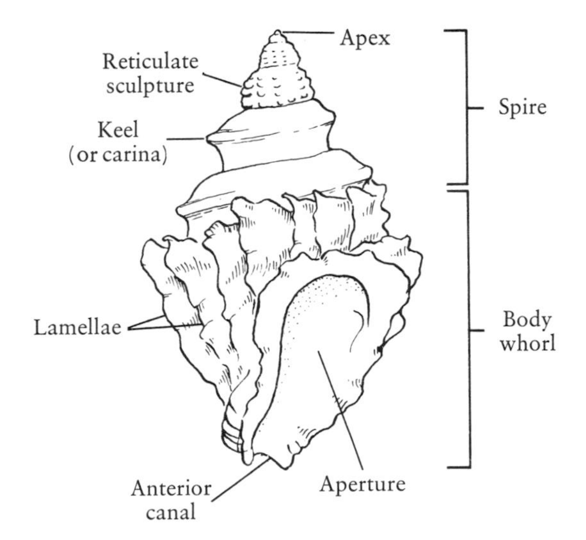
Apex
Reticulate sculpture
Keel (or carina)
Spire
Lamellae
Body whorl
Anterior canal
Aperture

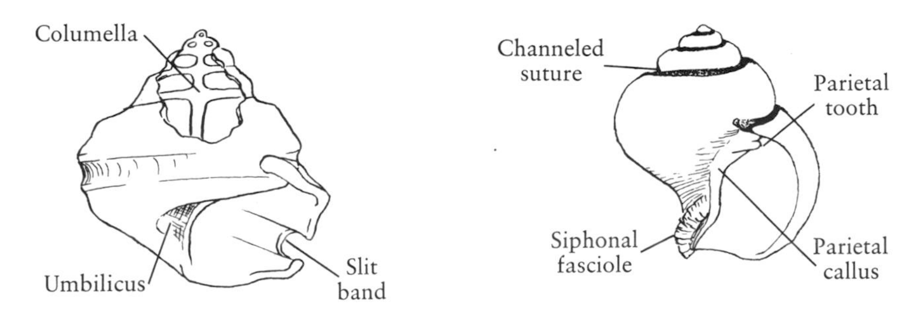
Columella
Umbilicus
Slit band
Channeled suture
Parietal tooth
Siphonal fasciole
Parietal callus

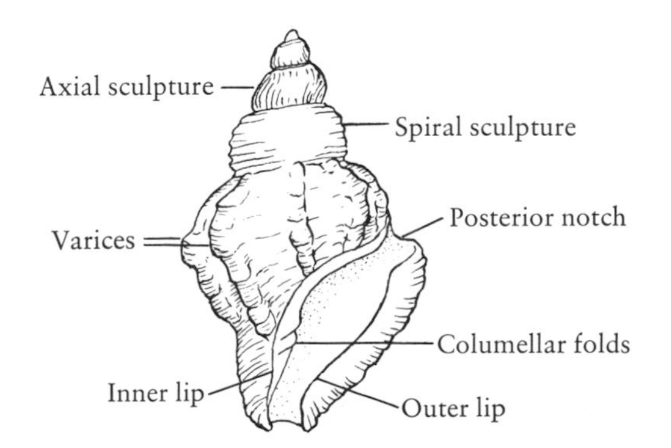
Axial sculpture
Spiral sculpture
Varices
Posterior notch
Columellar folds
Inner lip
Outer lip

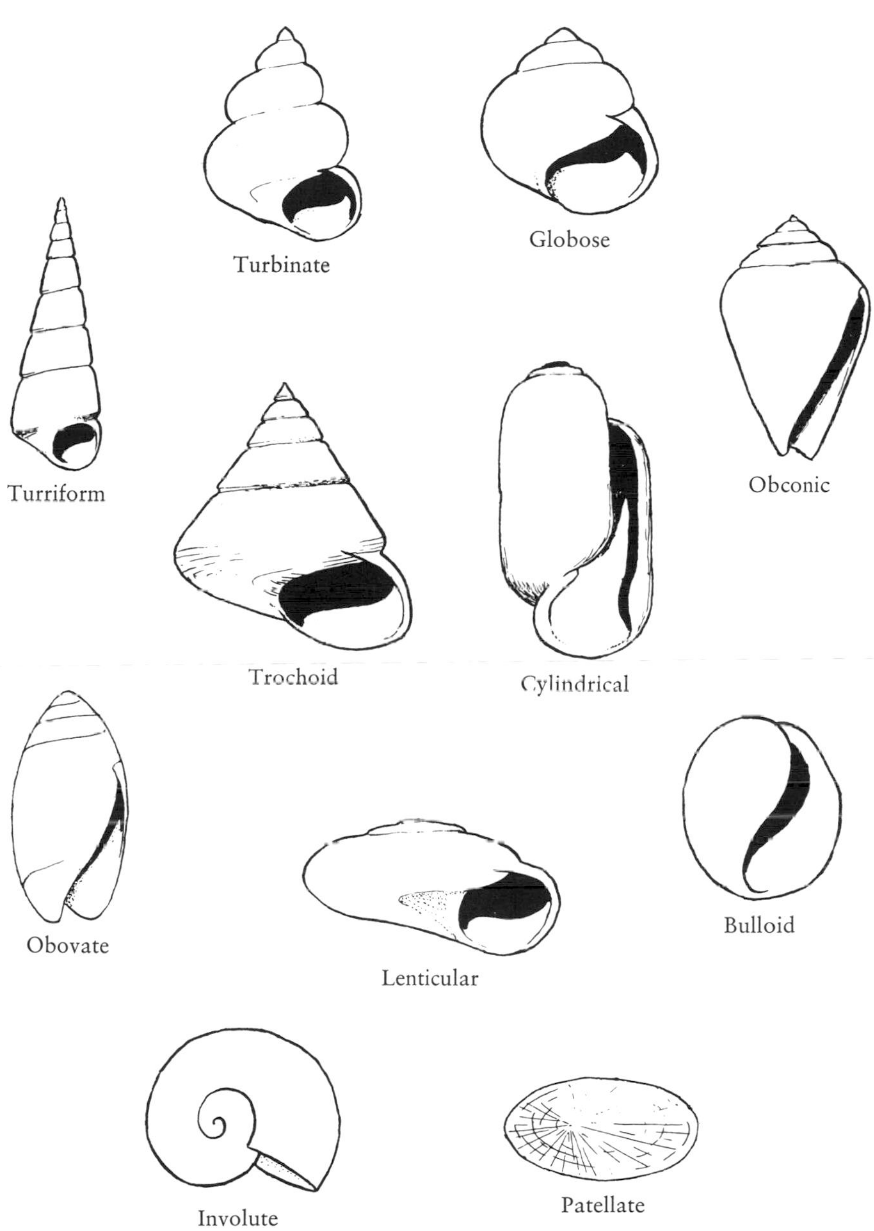
Turbinate
Globose
Turriform
Obconic
Trochoid
Cylindrical
Obovate
Bulloid
Lenticular
Involute
Patellate

SHELL FORMS

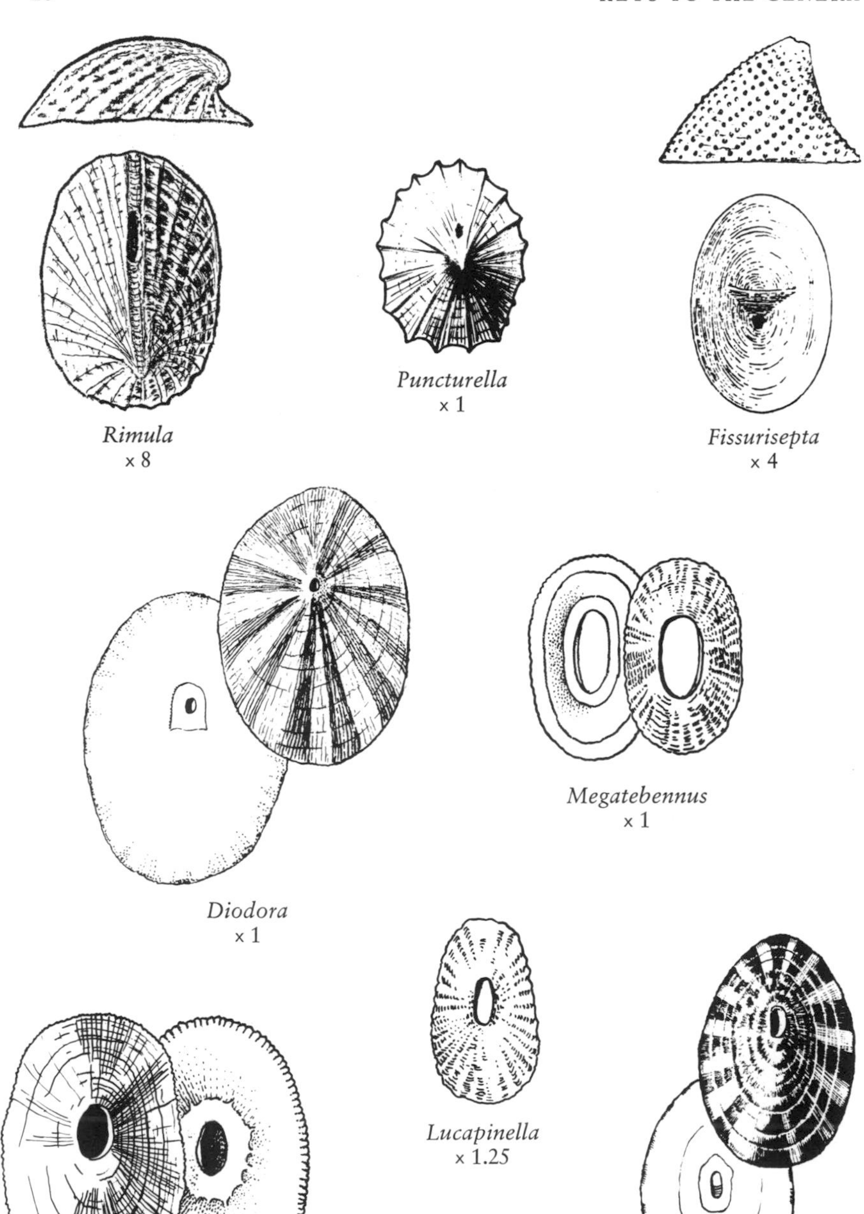

Rimula × 8

Puncturella × 1

Fissurisepta × 4

Diodora × 1

Megatebennus × 1

Lucapinella × 1.25

Megathura × 0.25

Fissurella × 1

1 Shell not spirally coiled—tubular, cap-shaped, or irregular in form 2

Shell spirally coiled for at least one complete turn (the coiling may be involute, the spire concealed by body whorl) 43

2(1) Shell patellate, i.e. low-conic (the underside open, revealing muscle scars) 3

Shell not patellate, i.e. high-conic to tubular (concealing muscle scars), petal-shaped, or irregular in form 27

3(2) With an opening at apex or on anterior slope of shell 4

Apex entire, no opening 11

4(3) Opening midway along anterior slope, elongate . . . . . *Rimula*

Opening at or near apex, rounded 5

5(4) With internal septum bridging apical cavity 6

Without internal septum; apical opening bordered by internal callus 7

6(5) Apical opening in front of a recurved apex, small . . . . . *Puncturella*

Opening at apex, small . . . . . *Fissurisepta*

7(5) Internal callus margin truncate (i.e. squarely cut off) posteriorly . . . . . *Diodora*

Internal callus of even width around orifice 8

8(7) Orifice widely oval, relatively large 9

Orifice narrowly oval, not relatively large 10

9(8) Inner margin set off by a broad shallow groove . . . . *Megatebennus*

Inner margin regularly crenate, not bordered by a groove . . . . . *Megathura*

10(8) Apical opening narrowed anteriorly; shell low . . . . . *Lucapinella*

Apical opening evenly oval; shell elevated . . . . . *Fissurella*

11(3) Interior with a deck or other shelly process 12

Interior with no secondary shelly process 15

12(11) Apex subcentral, shell tent-shaped 13

Apex near margin, not central; shell slipper-shaped 14

13(12) With an internal process suspended inside apex . . . . . *Crucibulum*

With a spiral deck, attached along outer margin . . . . . *Calyptraea*

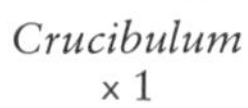

Crucibulum
× 1

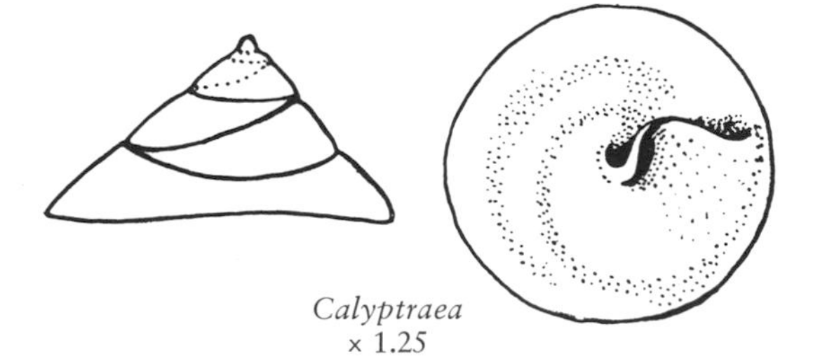

Calyptraea
× 1.25

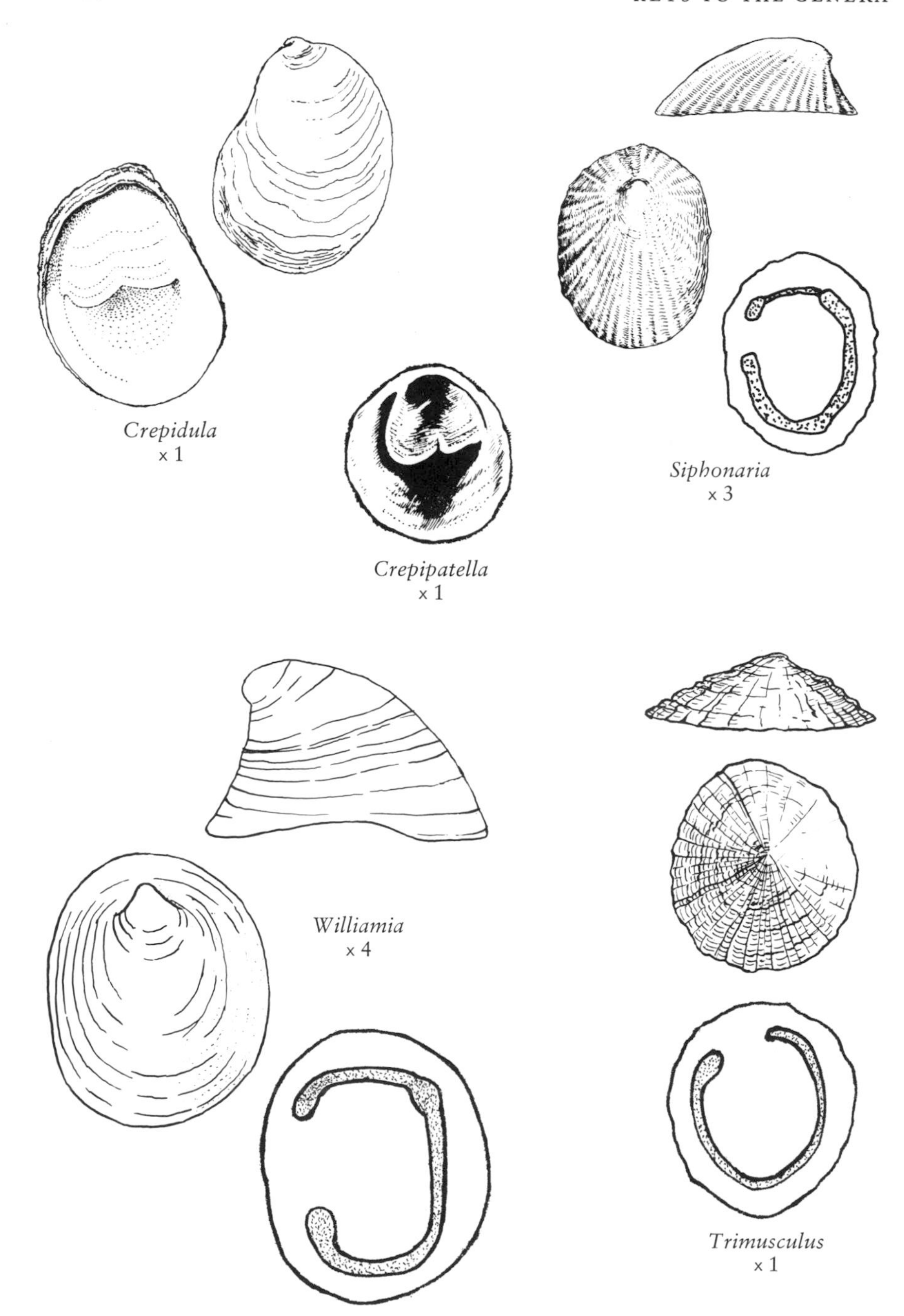
Crepidula
x 1
Crepipatella
x 1
Siphonaria
x 3
Williamia
x 4
Trimusculus
x 1

14(12) With a deck attached along both sides . . . . *Crepidula*
With a deck detached along one side . . . . *Crepipatella*
15(11) Muscle scar continuous across anterior end, interrupted on right (i.e. on left side in a ventral view) 16
Muscle scar interrupted anteriorly 17
16(15) Apex strongly hooked, shell smooth . . . . *Williamia*
Apex weakly hooked, shell irregularly sculptured . . . . . *Siphonaria*
17(15) Gap between ends of muscle scar excentric, slightly to right of midline (left in ventral view) . . . . *Trimusculus*
Gap anterior, centrally located 18
18(17) With an evident periostracum 19
Periostracum absent or thin 22
19(18) Periostracum membranous at margins, shell smooth . . . . *Tylodina*
Periostracum uniform throughout, shell usually sculptured 20
20(19) Shell thin, translucent, showing radial color stripes within . . . . *Piliscus*
Shell solid, not color-marked 21

Tylodina
× 1.5

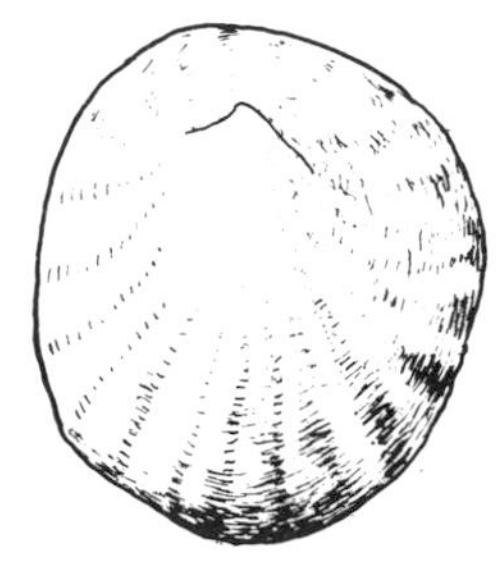

Piliscus
× 1.5

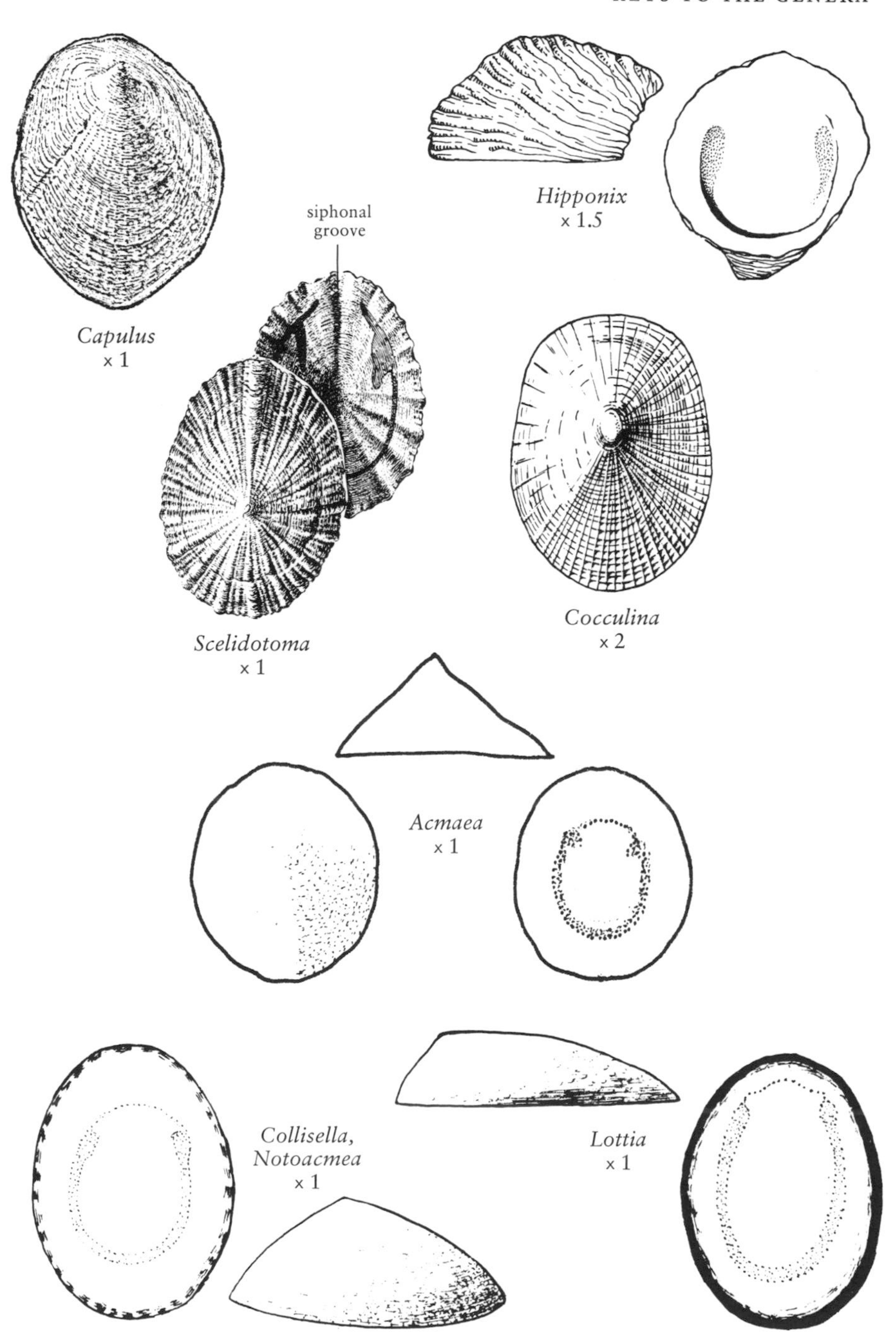
siphonal
groove
Capulus
× 1
Hipponix
× 1.5
Scelidotoma
× 1
Cocculina
× 2
Acmaea
× 1
Collisella,
Notoacmea
× 1
Lottia
× 1

21(20) Shell smooth or with a few radial striae .... *Capulus*
Sculpture of concentric lamellae somewhat intersected by fine radial riblets .... *Hipponix*

22(18) Anterior margin notched by the end of a raised siphonal groove .... *Scelidotoma*
Anterior margin evenly rounded, not notched 23

23(22) Apex turned backward; all species confined to deep water .... *Cocculina*
Apex turned forward, subcentrally located or toward anterior margin; not in deep water 24

24(23) Ends of horseshoe-shaped muscle scar joined by a thin, faint, even line 25
Ends of muscle scar joined by a distinctly curved line 26

25(24) Shell white to pink-rayed, apex nearly central .... *Acmaea*
Shell variously colored and marked (not white or pink-rayed), apex forward of center .... *Collisella, Notoacmea*
[*Collisella* having and *Notoacmea* lacking a pair of uncini on basal plate of radula]

26(24) Line joining ends of muscle scar irregularly sinuous; internal margin of shell dark-colored .... *Lottia*
Line joining ends of muscle scar regularly curved toward margin; internal border white .... *Cryptobranchia, Lepeta*
[Fused inner lateral teeth of radula elongate and pointed in *Lepeta*, rectangular in *Cryptobranchia*]

27(2) Shell internal, petal-shaped 28
Shell external, high-conic to tubular or irregular in form 33

28(27) With a minute spiral initial whorl 29
Without a spiral initial whorl 31

Cryptobranchia, Lepeta
× 1.5

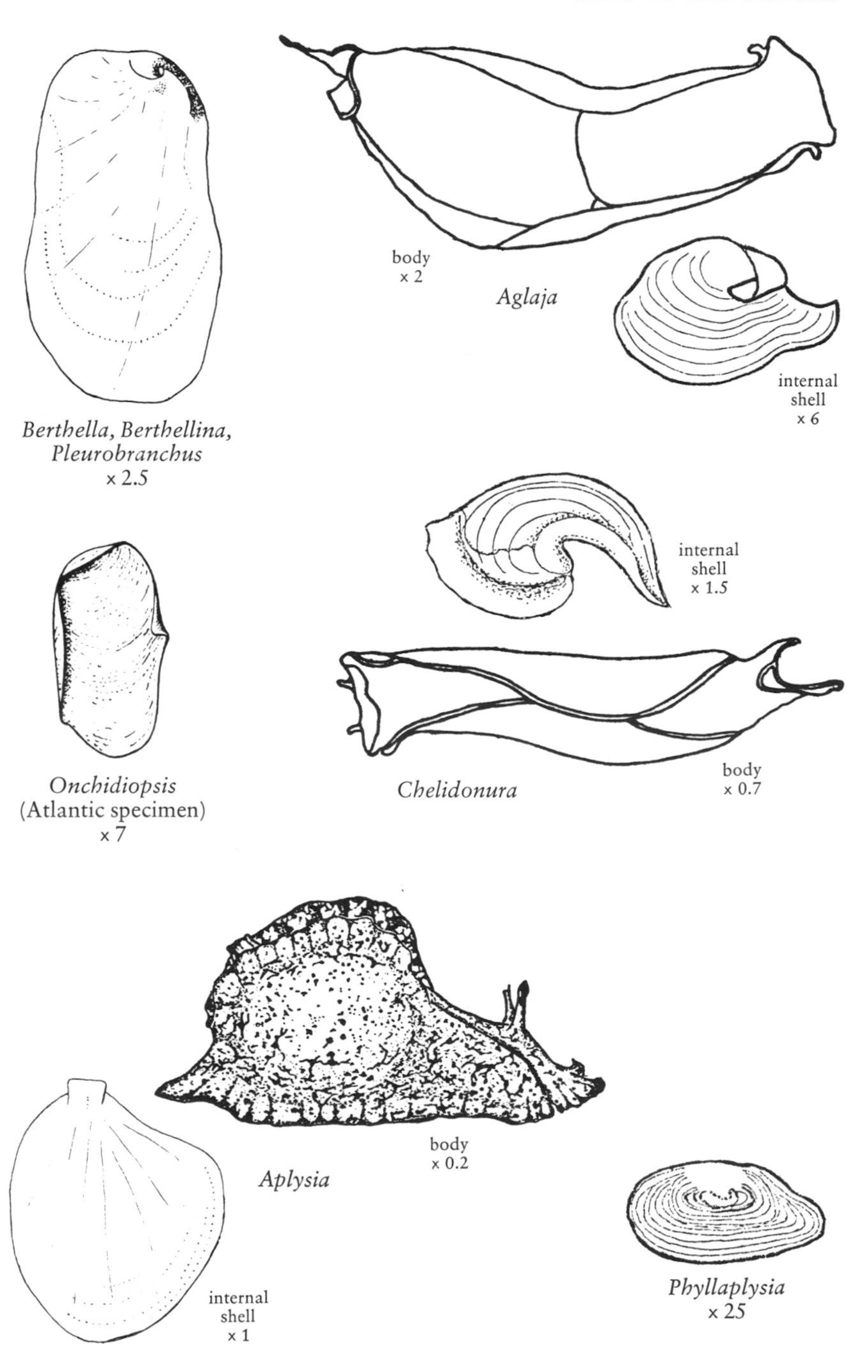

Berthella, Berthellina, Pleurobranchus x 2.5

Aglaja

Onchidiopsis (Atlantic specimen) x 7

Chelidonura

Aplysia

Phyllaplysia x 25

29(28) Shell subquadrate in outline .... *Pleurobranchus, Berthellina, Berthella*
[*Pleurobranchus* with a pedal gland posteriorly, gill rhachis granulate; *Berthella* without pedal gland, gill rhachis smooth, radular teeth short, hooked, smooth; *Berthellina* like *Berthella*, but radular teeth bladelike, posterior edges serrate]
Shell elongate or somewhat spirally twisted 30

30(29) Inner side of shell with a spiral process or columella...... *Aglaja*
Inner side nearly as flat as outer surface .... *Chelidonura*

31(28) Margins irregularly sinuous .... *Onchidiopsis*
Marginal sinuosity even or interrupted on one side by a knoblike area 32

32(31) Shell invariably present in mantle cavity, its nucleus marginal, concealed by knoblike lamella .... *Aplysia*
Shell not invariably present, lacking a marginal lamella .... *Phyllaplysia*

33(27) Shell inflated; aperture somewhat obstructed 34
Shell not inflated; aperture not obstructed; mostly tubular in outline 35

34(33) Lips of aperture thin and sharp .... *Cavolinia*
Lips of aperture with a rolled edge .... *Diacria*

Cavolinia
x 4

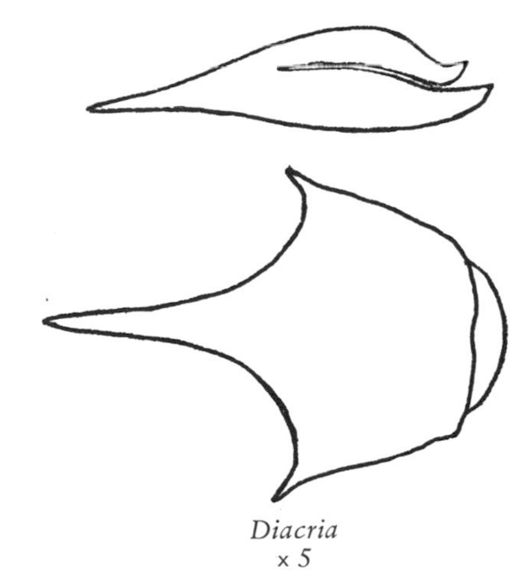

Diacria
x 5

Fartulum
x 13

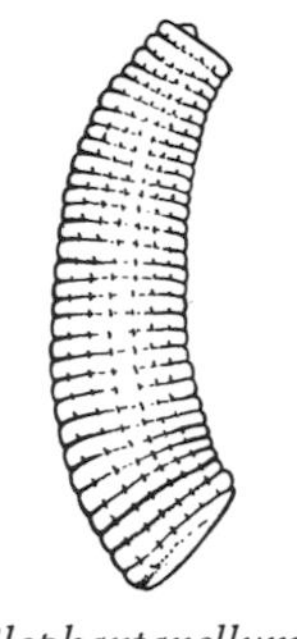

Elephantanellum
x 9

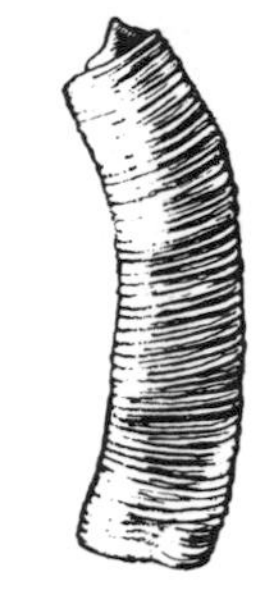

Micranellum
x 8

Caecum
x 15

Cuvierina
x 5

Hyalocylis
x 4

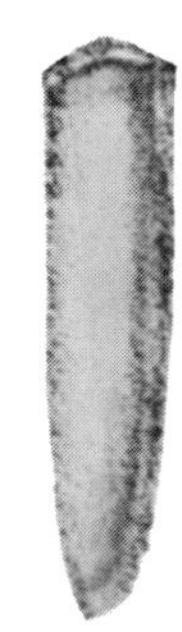

Creseis
x 6

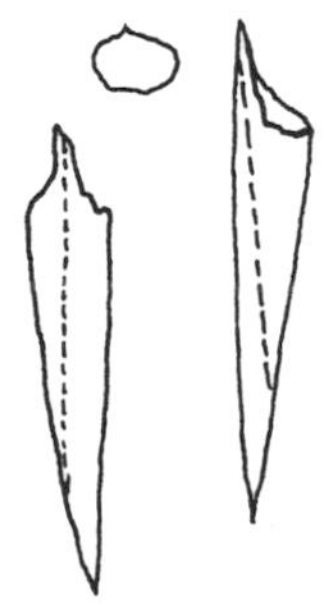

Styliola
x 3

35(33) Shell with an apical plug; shell texture porcelaneous 36
Shell with evenly pointed apex, not ending in a plug; texture thin, almost transparent 39
36(35) Shell smooth, mostly brown in color . . . . *Fartulum*
Shell with concentric annulations, mostly white 37
37(36) With annular rings and longitudinal ridges . . . . *Elephantanellum*
With annular rings only 38
38(37) With rings closely spaced . . . . *Micranellum*
With rings distantly spaced . . . . *Caecum*
39(35) Posterior end bluntly rounded . . . . *Cuvierina*
Posterior end acutely pointed 40
40(39) Surface with raised transverse annulations . . . . *Hyalocylis*
Surface smooth or with longitudinal ribs .41
41(40) Smooth, cross-section nearly circular; apertural margin even . . . . *Creseis*
With one or more longitudinal ribs; apertural margin prolonged dorsally 42
42(41) Longitudinal ribs weak, apertural margins without lateral prolongations or spurs . . . . *Styliola*
Longitudinal ribs well developed, forming spurs on either side of aperture . . . . *Clio*
43(1) Aperture entire, with no anterior canal or notch 44
Aperture with an anterior notch or canal 164
44(43) Shell becoming uncoiled with growth 45
Shell not becoming uncoiled with growth 49
45(44) Early whorls of post-juvenile shell regularly coiled, later whorls becoming detached; shell weakly cemented to substrate. . . . . *Vermicularia*
Early whorls of post-juvenile shell irregular; shell firmly cemented to rocks or other shells 46

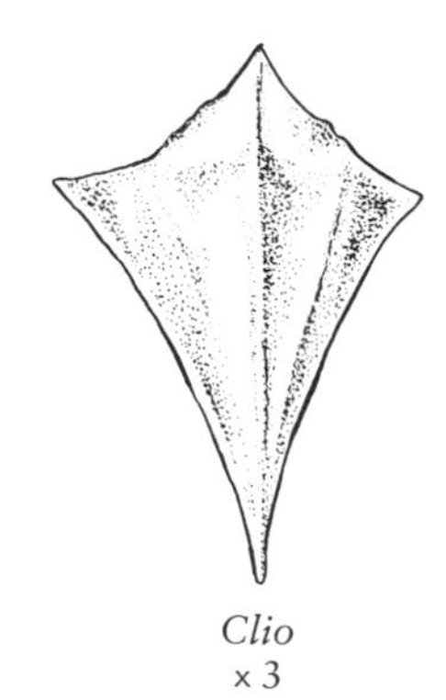
Clio
x 3

Vermicularia
x 0.75

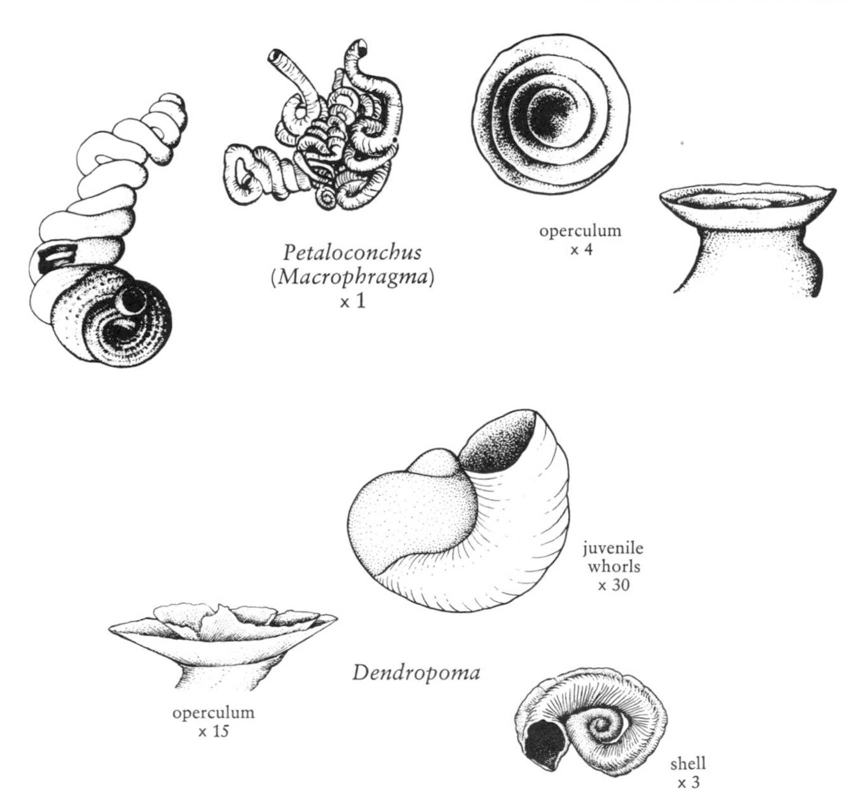

Petaloconchus
(*Macrophragma*)
x 1

Dendropoma

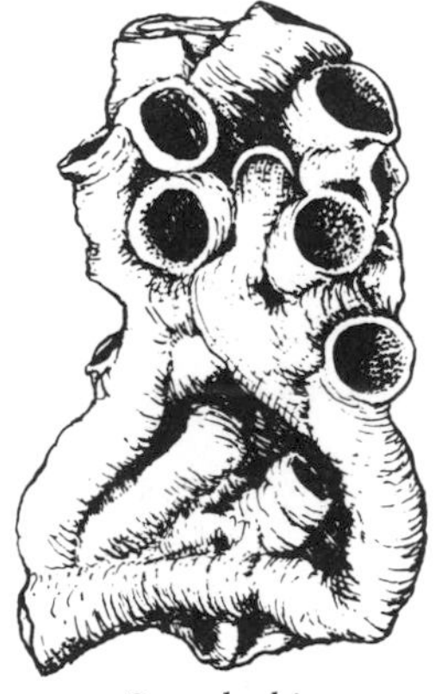

Serpulorbis
x 2

Vermetus
(*Thylaeodus*)
x 5

operculum
x 20

46(45) Medial whorls with internal laminae, showing on the columella in broken sections . . . . *Petaloconchus (Macrophragma)*
Shell without internal structure on columella 47
47(46) Shell partially buried in an eroded channel; operculum present, as large in diameter as the aperture . . . . *Dendropoma*
Shell attached to surface, not partially immersed; operculum, if present, smaller than aperture 48
48(47) Adult tubes more than 4 mm in diameter, in dense and contorted masses; operculum wanting . . . . *Serpulorbis*
Adult tubes less than 3 mm in diameter, solitary or in small colonies; operculum one-half as wide as aperture . . . . *Vermetus (Thylaeodus)*
49(44) Coiling involute; shells small in proportion to size of animal; habitat pelagic 50
Coiling helical; shells usually large enough to cover retracted animal; habitat benthonic 55
50(49) Shell laterally compressed, double-keeled . . . . *Carinaria*
Shell not laterally compressed nor double-keeled 51
51(50) Shell more or less internal, not keeled 52
Shell not internal, with single evident keel 53
52(51) Margin of aperture produced as winglets; shell partially concealed by soft parts . . . . *Cardiapoda*
Margin of aperture even, thin; shell entirely concealed by soft parts . . . . *Gastropteron*

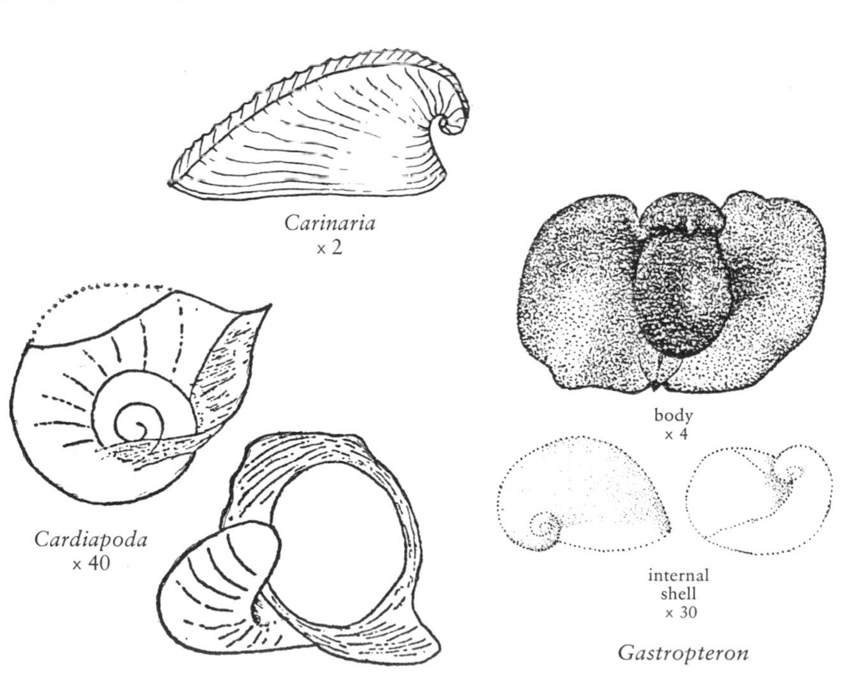

Carinaria x 2

Cardiapoda x 40

Gastropteron

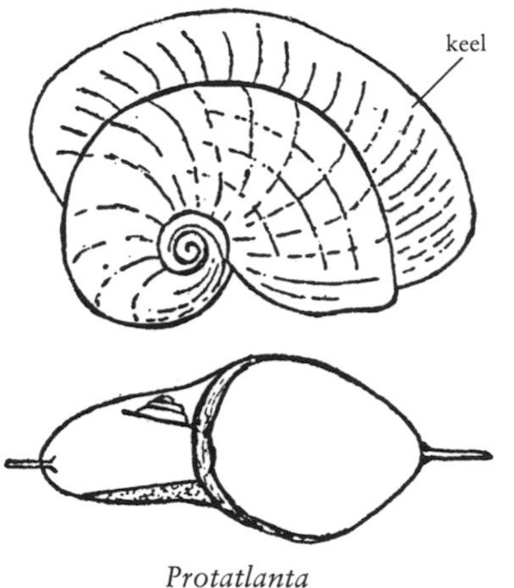

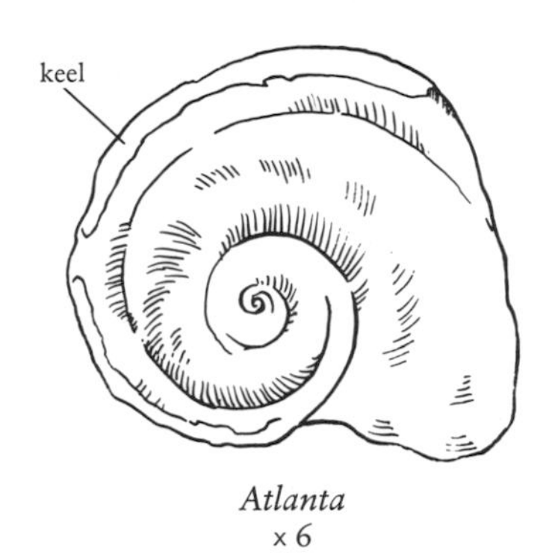

Atlanta
x 6

Protatlanta
x 20

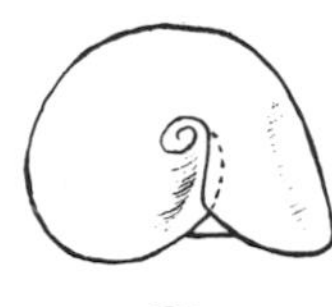

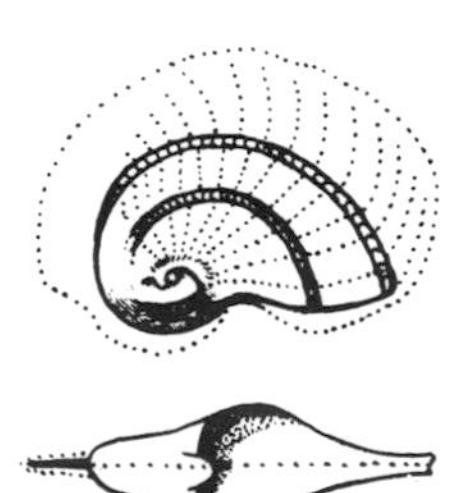

Oxygyrus
(Atlantic specimen)
x 1.5

Philine
x 7

Bulla
x 1

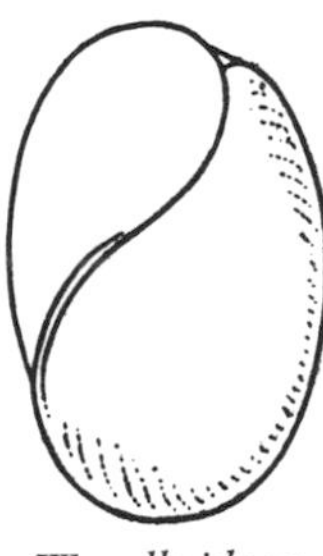

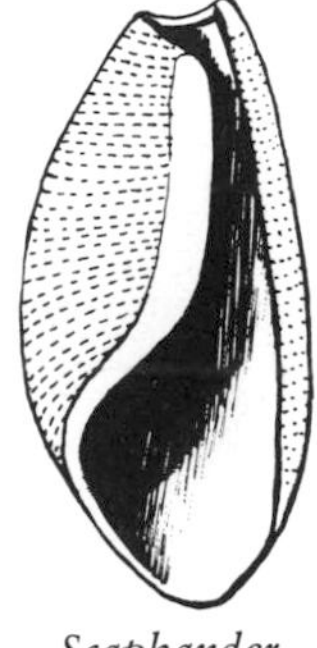

Woodbridgea
x 20

Scaphander
x 3

53(51) Planorboid, spire nearly concealed; keel cartilaginous . . . . . . *Oxygyrus*
Not planorboid, spire somewhat projecting; keel chalky to cartilaginous 54
54(53) Keel continuing up to shell aperture, cartilaginous . . . . . *Protatlanta*
Keel sloping away or narrowing near aperture, mostly chalky . . . . . *Atlanta*
55(49) Shell bulloid (i.e. spire concealed, aperture as long as shell) 56
Shell not bulloid 60
56(55) Shell extremely thin, almost membranous 57
Shell thin but not extremely so 58
57(56) Surface smooth . . . . . *Philine*
Surface with spiral striae or punctations . . . . . *?Woodbridgea*
58(56) Aperture evenly arched, not flaring below . . . . . *Bulla*
Aperture flaring below 59
59(58) Surface spirally punctate; outer lip carinate at apex . . . *Scaphander*
Surface smooth; outer lip not carinate at apex . . . . . *Haminoea*
60(55) Shell cylindrical 61
Shell not cylindrical 69
61(60) With several columellar folds 62
With not more than one columellar fold 63
62(61) Adults larger than 5 mm in height; shells with color bands . . . *Volvarina*
Adults smaller than 4 mm in height; shells white . . . . . *Cystiscus**

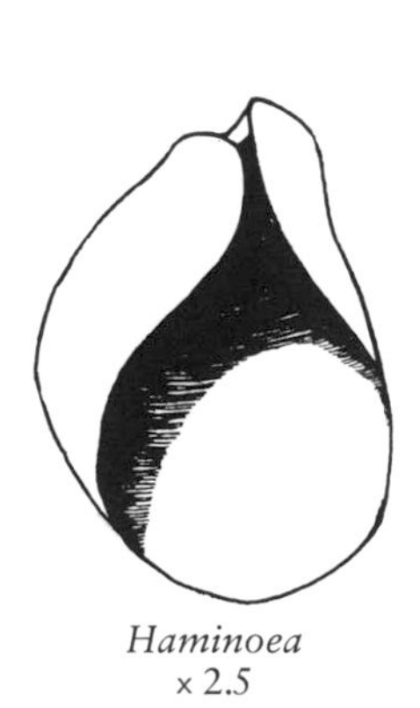

Haminoea
x 2.5

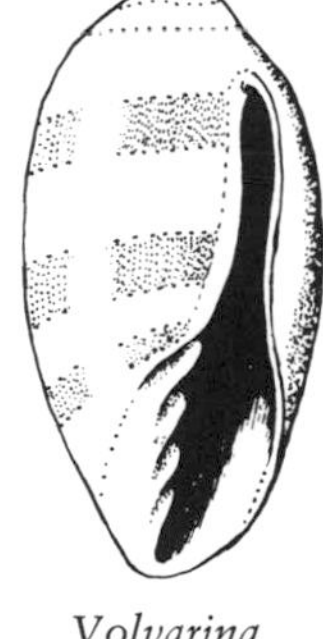

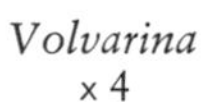

Volvarina
x 4

Cystiscus
x 15

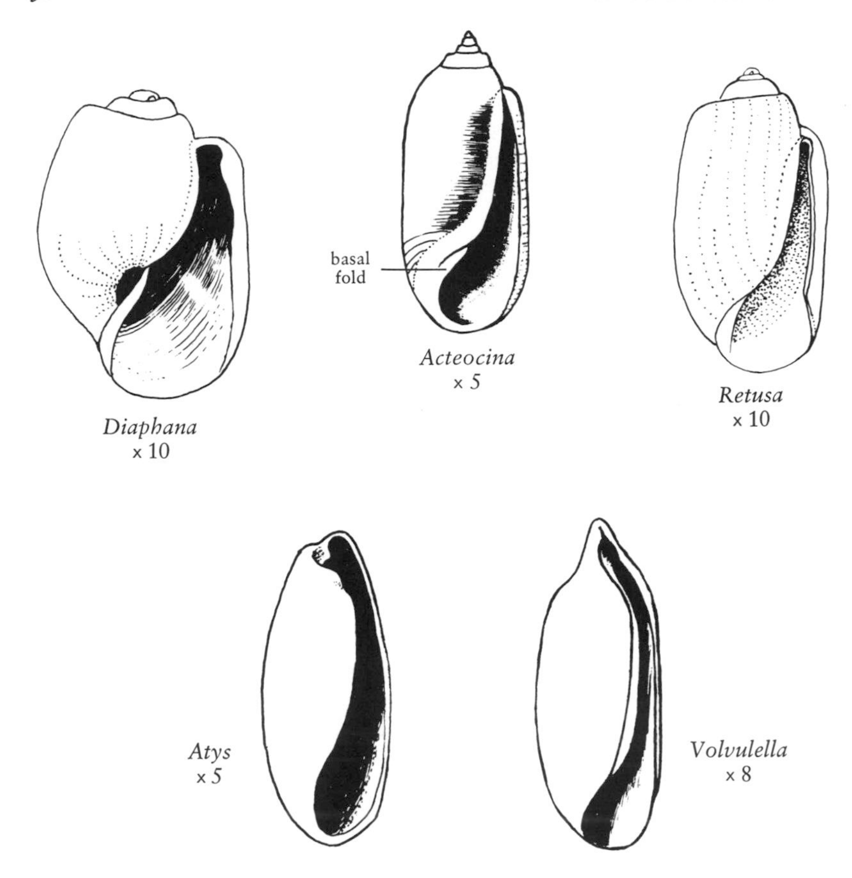

Cylichna
x 5

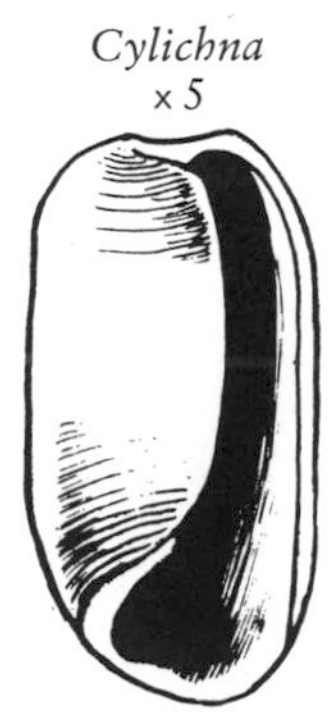

Haliotis
x 0.3

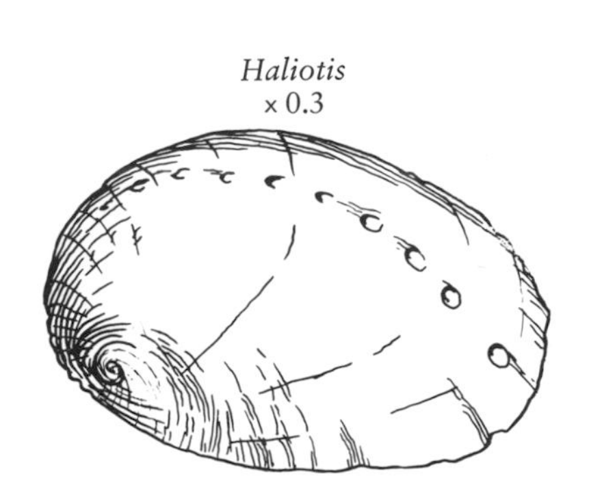

Sulcoretusa
x 10

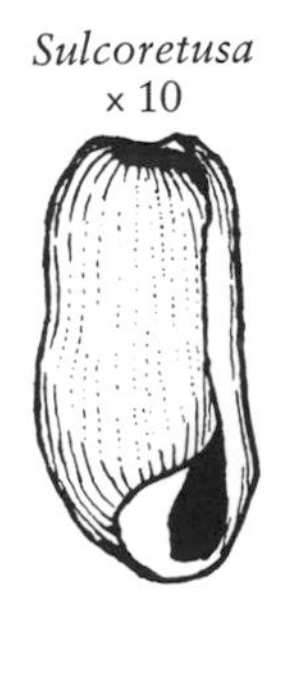

63(61) Spire apparent, from slightly to moderately elevated 64
Spire sunken or concealed 66
64(63) Aperture somewhat flaring anteriorly .... *Diaphana*
Aperture not flaring anteriorly 65
65(64) Columella with a basal fold .... *Acteocina*
Columella smooth .... *Retusa*
66(63) Spire concealed by apex 67
Spire sunken but not concealed 68
67(66) Outer lip twisted into a low fold above spire .... *Atys*
Outer lip not twisted but prolonged above spire into a spine-like extension .... *Volvulella*
68(66) Sculpture faint, of incised spiral lines .... *Cylichna*
Sculpture moderately developed, of axial striae .... *Sulcoretusa*
69(60) Shell with a peripheral sinus, a slit band, or a row of holes 70
Without peripheral sinus, slit band, or holes 74
70(69) Shell ear-shaped; body whorl with a row of holes .... *Haliotis*
Shell not ear-shaped; body whorl with a peripheral sinus and not more than one hole 71
71(70) Color light to dark purple .... *Janthina*
Color white 72
72(71) Outer lip with a single elongate perforation closed at margin .... *Sinezona*
Outer lip with an open slit 73

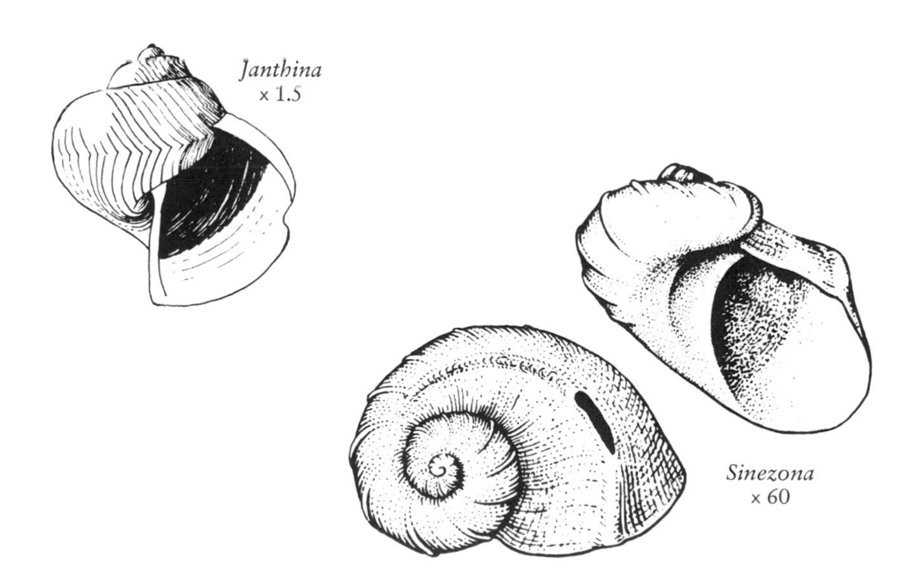
Janthina x 1.5

Sinezona x 60

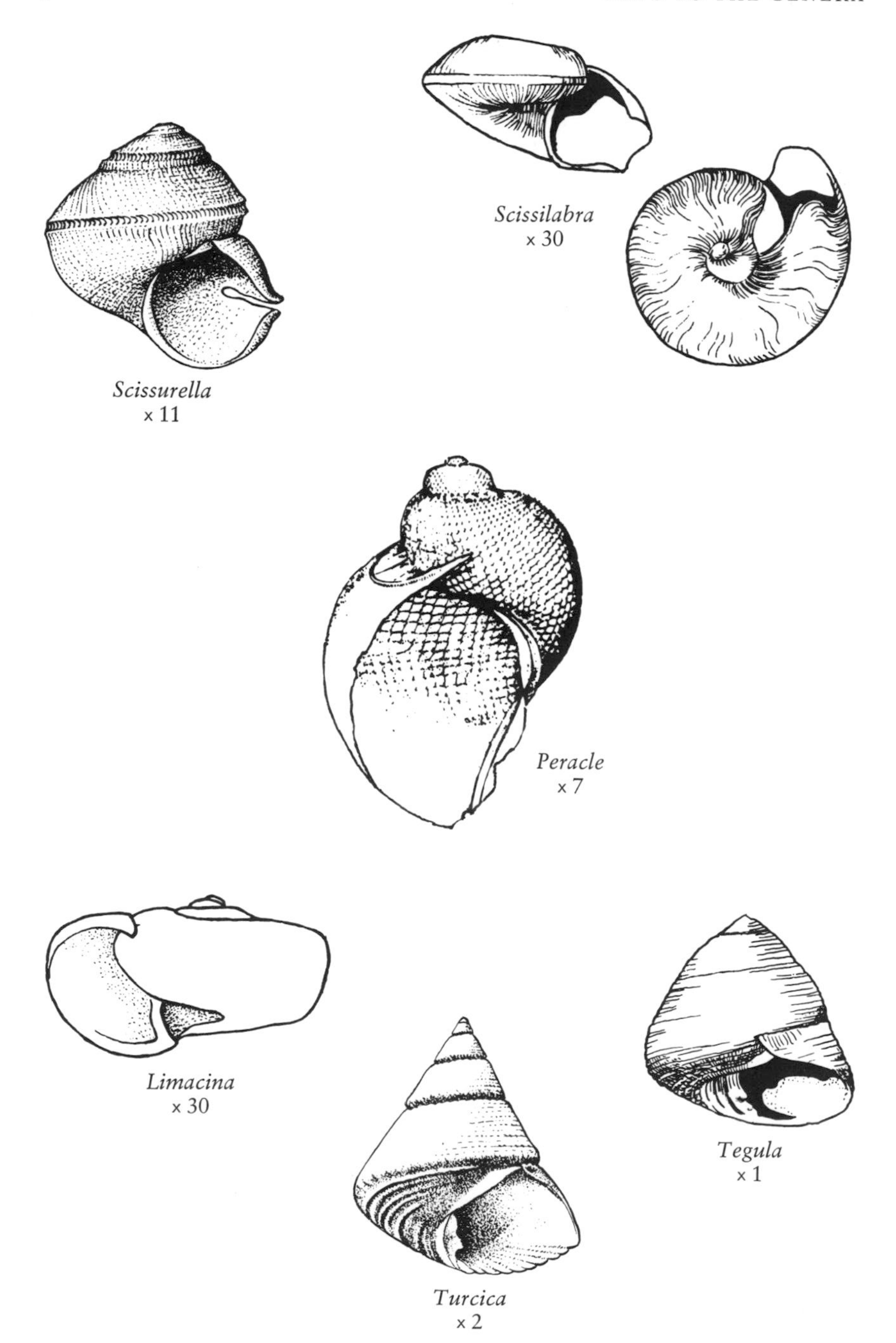

Scissilabra
× 30

Scissurella
× 11

Peracle
× 7

Limacina
× 30

Tegula
× 1

Turcica
× 2

73(72) Shell turbinate or trochoid in outline, spire evident ..... *Scissurella*
Shell lenticular, spire depressed .... *Scissilabra*
74(69) Coiling sinistral 75
Coiling dextral 76
75(74) Shell surface smooth .... *Limacina*
Shell surface covered with a membranous hexagonal network .... *Peracle*
76(74) Interior nacreous 77
Interior porcelaneous 92
77(76) Columella with nodes or denticles 78
Columella smooth 80
78(77) Outline conical or trochoid .... *Turcica*
Outline inflated to globose 79
79(78) Denticles strong, operculum thin, multispiral .... *Tegula*
Denticles weak, operculum calcareous, paucispiral .. *Homalopoma*
80(77) Shell large, solid, thick 81
Shell small to medium-sized, not heavy and solid 82
81(80) Periphery angulate, base nonumbilicate, operculum calcareous .... *Astraea*
Periphery rounded, base umbilicate, operculum not calcareous .... *Norrisia*

Homalopoma
x 4

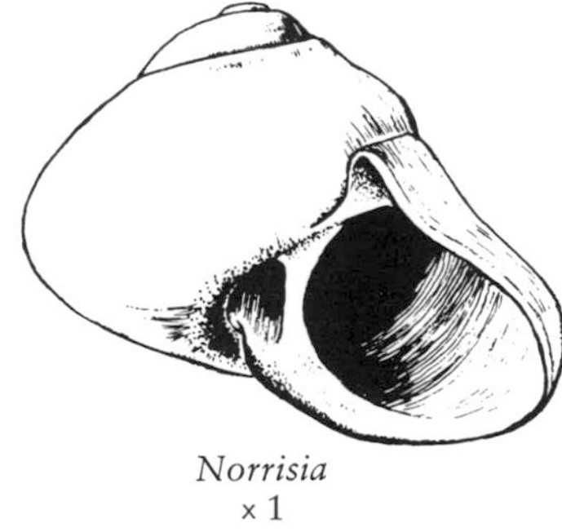

Norrisia
x 1

Astraea
x 0.8

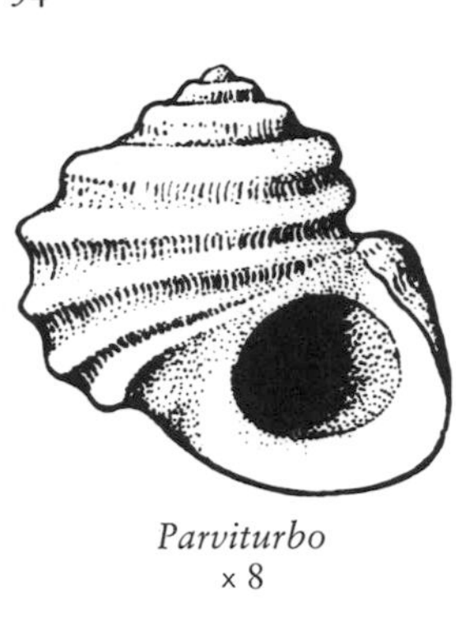

Parviturbo
× 8

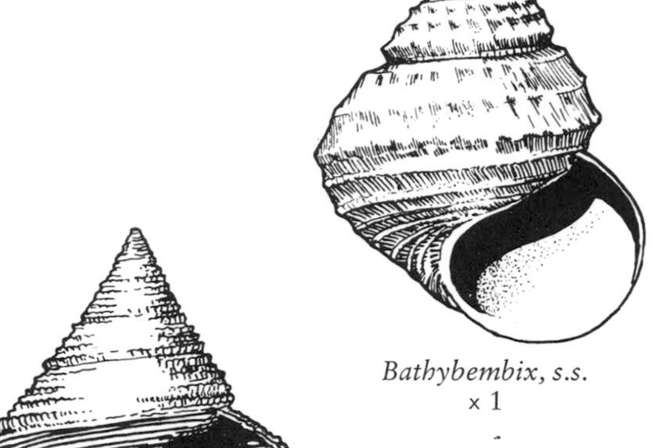

Bathybembix, s.s.
× 1

Calliostoma
× 1.5

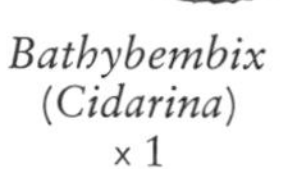

Bathybembix
(*Cidarina*)
× 1

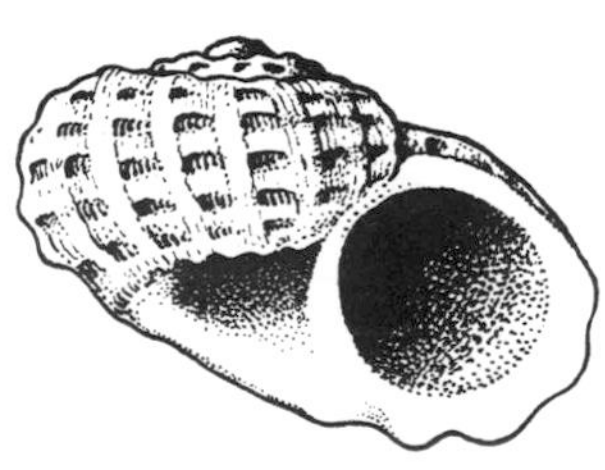

Liotia
× 12

Calliotropis
× 2

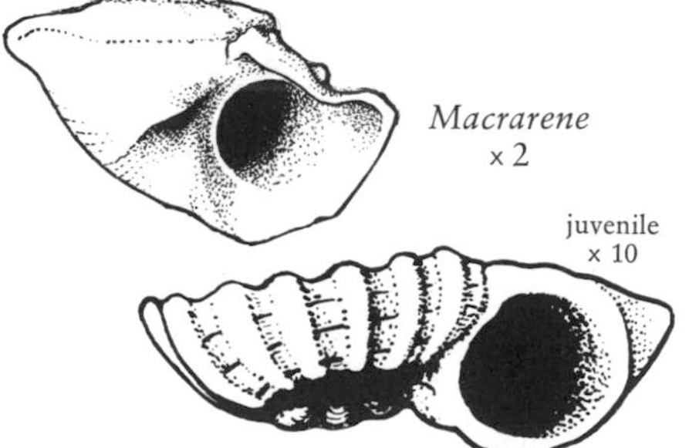

Macrarene
× 2

juvenile
× 10

Solariella
× 2

82(80) Base nonumbilicate 83
Base umbilicate 86
83(82) Sculpture of a few large cords, axials weak . . . . *Parviturbo*
Sculpture of numerous spiral ribs, axials evident to strong 84
84(83) Outline conical to trochoid . . . . *Calliostoma*
Outline angulate to tabulate 85
85(84) Shell thin, periostracum tightly adherent . . . . *Bathybembix, s.s.*
Shell sturdy, periostracum fibrous . . . . *Bathybembix (Cidarina)*
86(82) Axial sculpture stronger than growth lines 87
Axial sculpture wanting or of growth lines only 90
87(86) Sculpture evenly reticulate . . . . *Liotia*
Sculpture not of evenly reticulate ribbing 88
88(87) Axial ribs few, large; umbilicus not set off . . . . *Macrarene*
Axial ribs numerous, small; umbilicus bordered by a spiral thread 89
89(88) Spire with a few strong beaded cords; radula with lateral teeth numerous . . . . *Calliotropis*
Spire with fine, nearly smooth spiral cords; radula with lateral teeth reduced in number . . . . *Solariella*
90(86) Small (diameter less than 6 mm), tending toward variegated color patterns; radula with unusually few marginal teeth . . . . *Lirularia*
Medium-sized (diameter 6 to 20 mm), unicolored, radula with normally numerous marginal teeth of rhipidoglossate form . . . 91
91(90) Surface smooth, lustrous . . . . *Margarites, s.s.*
Surface sculptured with spiral riblets . . . . . . *Margarites (Pupillaria)*

Lirularia
×6

Margarites (Pupillaria)
×2.5

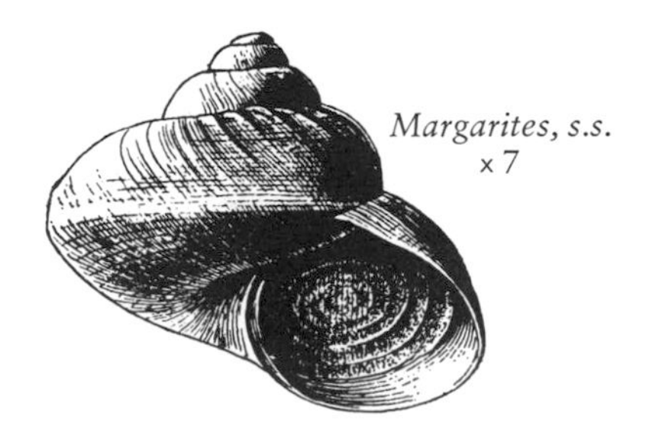
Margarites, s.s.
×7

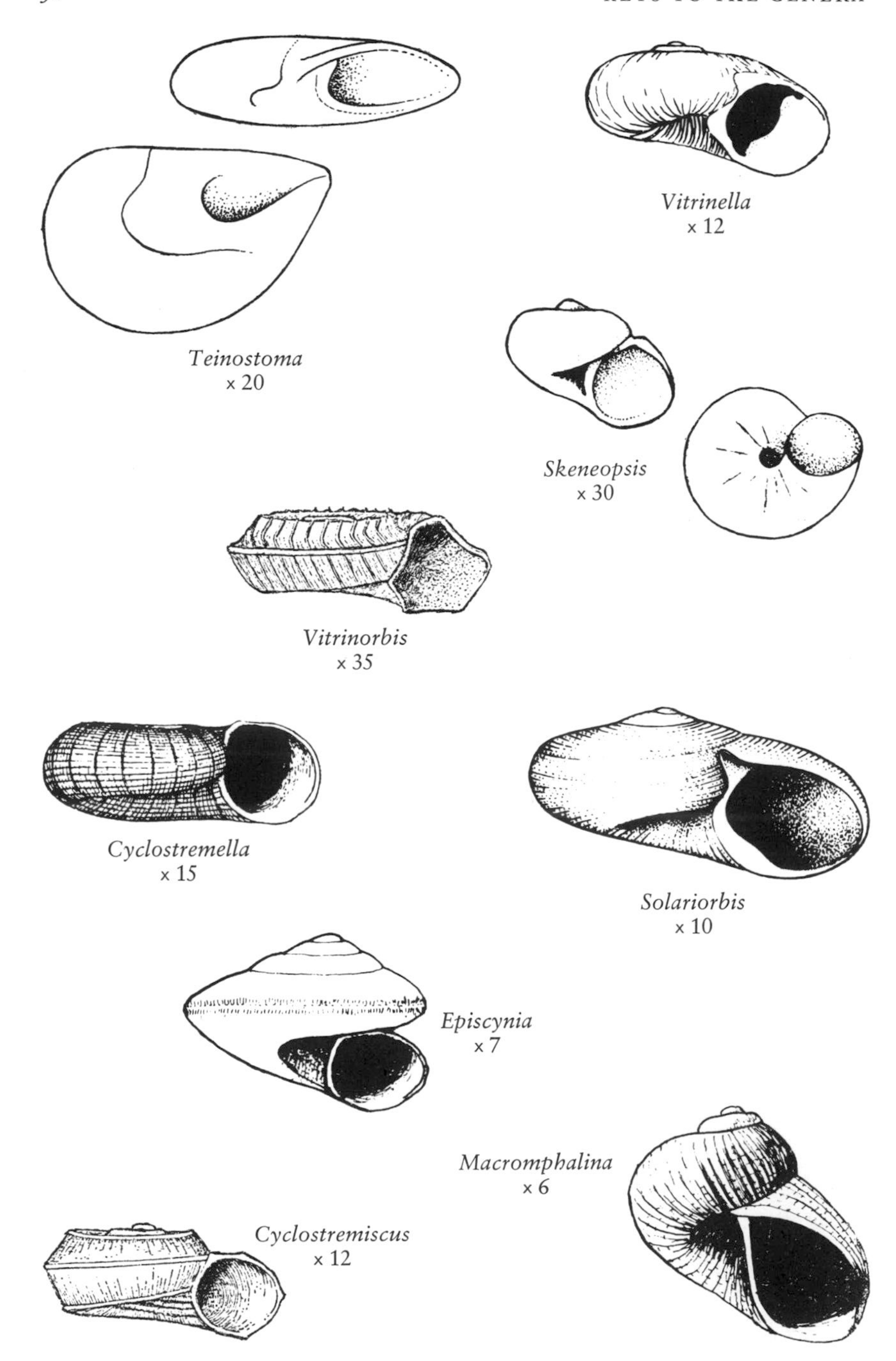
Teinostoma
x 20
Vitrinella
x 12
Skeneopsis
x 30
Vitrinorbis
x 35
Cyclostremella
x 15
Solariorbis
x 10
Episcynia
x 7
Macromphalina
x 6
Cyclostremiscus
x 12

92(76) Diameter greater than height 93
Diameter equal to or less than height 108
93(92) Small to minute shells, diameter less than 6 mm 94
Medium-sized shells, diameter 6 mm or more 104
94(93) Base of shell with a conspicuous callus pad .... *Teinostoma*
Base of shell umbilicate, with no callus covering 95
95(94) Surface smooth or nearly so 96
Surface sculptured, either striate or carinate 97
96(95) Umbilicus wide, shallow .... *Vitrinella*
Umbilicus narrow, deep .... *Skeneopsis*
97(95) Coiling planorboid, spire depressed 98
Coiling helical, spire low but evident 99
98(97) Periphery angulate, spire flat .... *Vitrinorbis*
Periphery rounded, spire sunken .... *Cyclostremella*
99(97) Columellar lip with a thickened lobe of callus .... *Solariorbis*
Columellar lip of even width throughout 100
100(99) Shell spirally keeled, with one or more carinae 101
Shell with rounded periphery 102
101(100) Body whorl with a single keel; outer lip of body whorl attached below peripheral carination of previous whorl .... *Episcynia*
Body whorl with more than one keel; suture concealing peripheral angulations of previous whorls .... *Cyclostremiscus*
102(100) Sculpture both axial and spiral but subdued .... *Macromphalina*
Sculpture of fine spiral threads only 103
103(102) Adult shell with a single postnuclear whorl; operculum thin, noncalcareous .... *Leptogyra*
Shell with two or more postnuclear whorls; operculum calcareous .... *Moelleria*
104(93) With a conspicuous periostracum 105
Without any evident periostracum 107
105(104) Periostracum smooth, glossy .... *Haloconcha*
Periostracum velvety 106

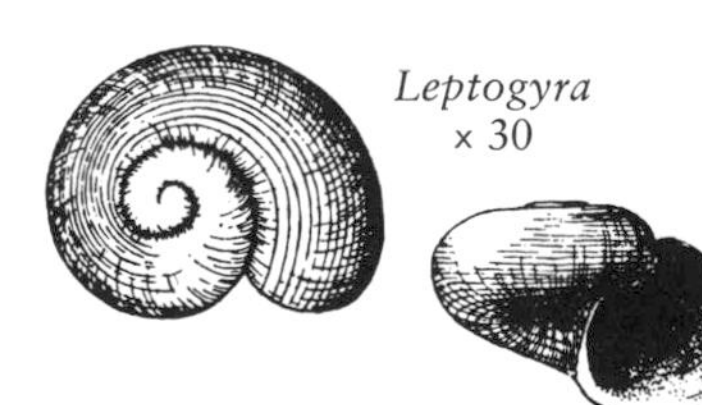
Leptogyra
x 30

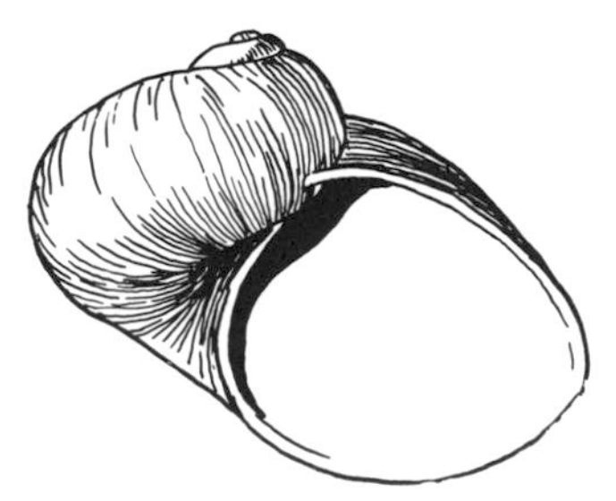
Haloconcha
x 8

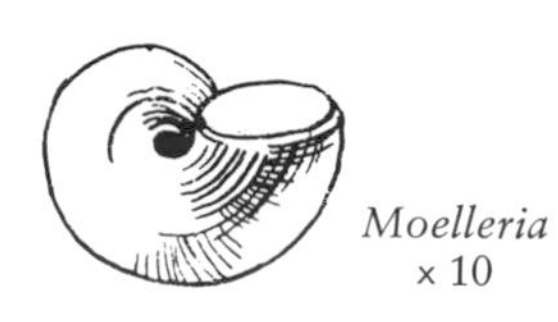
Moelleria
x 10

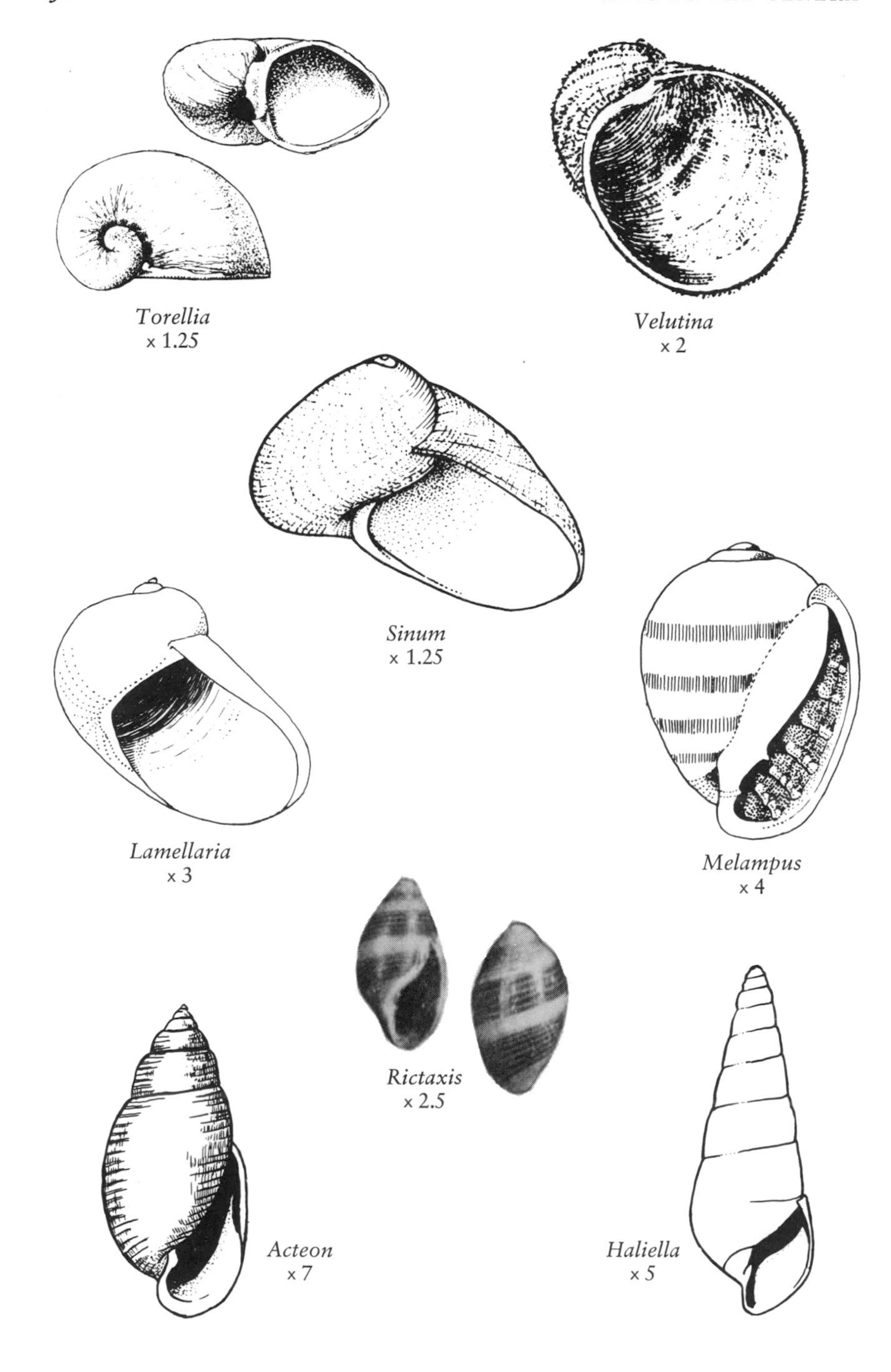

Torellia × 1.25

Velutina × 2

Sinum × 1.25

Lamellaria × 3

Melampus × 4

Rictaxis × 2.5

Acteon × 7

Haliella × 5

106(105) Body whorl incompletely calcified; periostracum leathery, flexible . . . . . . *Torellia*
Shell thin but well calcified; periostracum somewhat fringed or fuzzy . . . . . . *Velutina*

107(104) Surface with fine, regular spiral riblets . . . . . . *Sinum*
Surface smooth or with irregular malleations . . . . . . *Lamellaria*

108(92) Columella with one or more folds, or inner lip with a strong spiral ridge entering aperture 109
Columella and inner lip smooth 120

109(108) Outer lip with raised lirae within . . . . . . *Melampus*
Outer lip not marked with raised lirae 110

110(109) Interspaces between spiral ribs punctate or pitted . . . . . . *Acteon, Rictaxis*
[*Acteon* with numerous fine radular teeth of uniform size in each row; *Rictaxis* with five lateral radular teeth per row, one larger than the rest]
Spiral sculpture without punctations 111

111(110) Columella with no more than one fold 112
Columella with two or more folds 115

112(111) Nuclear whorls with normal dextral coiling, tapering, not immersed in later spire whorls . . . . . . *Haliella*
Nuclear whorls with heterostrophic coiling, somewhat immersed in first spire whorls 113

113(112) Columella smooth below but bordered above by a spiral rib that enters the aperture . . . . . . *Peristichia*
Columella with one fold or low ridge 114

114(113) Columella with one clear-cut, narrow fold . . . . . . *Odostomia*
Columella with a broad, low spiral ridge or swelling . . . . . . *Iselica*

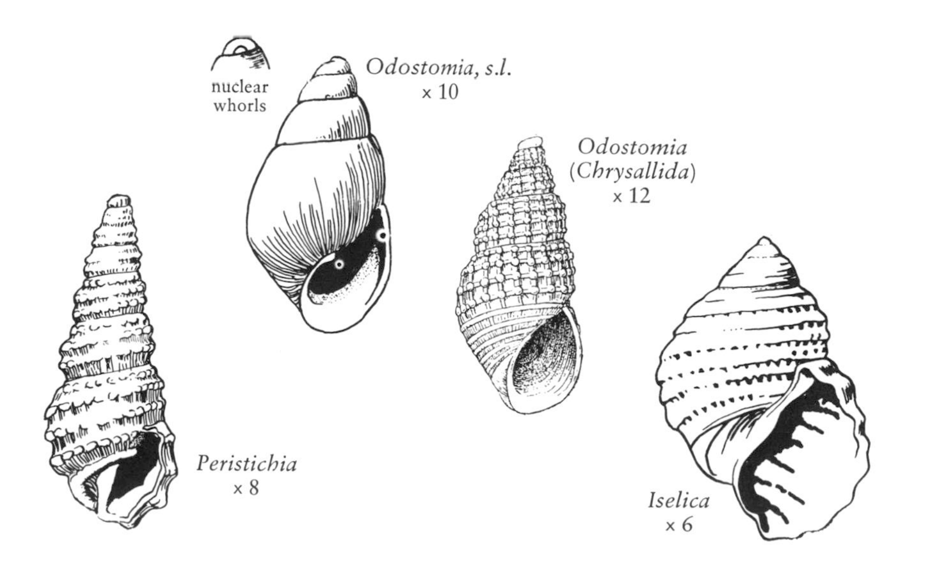

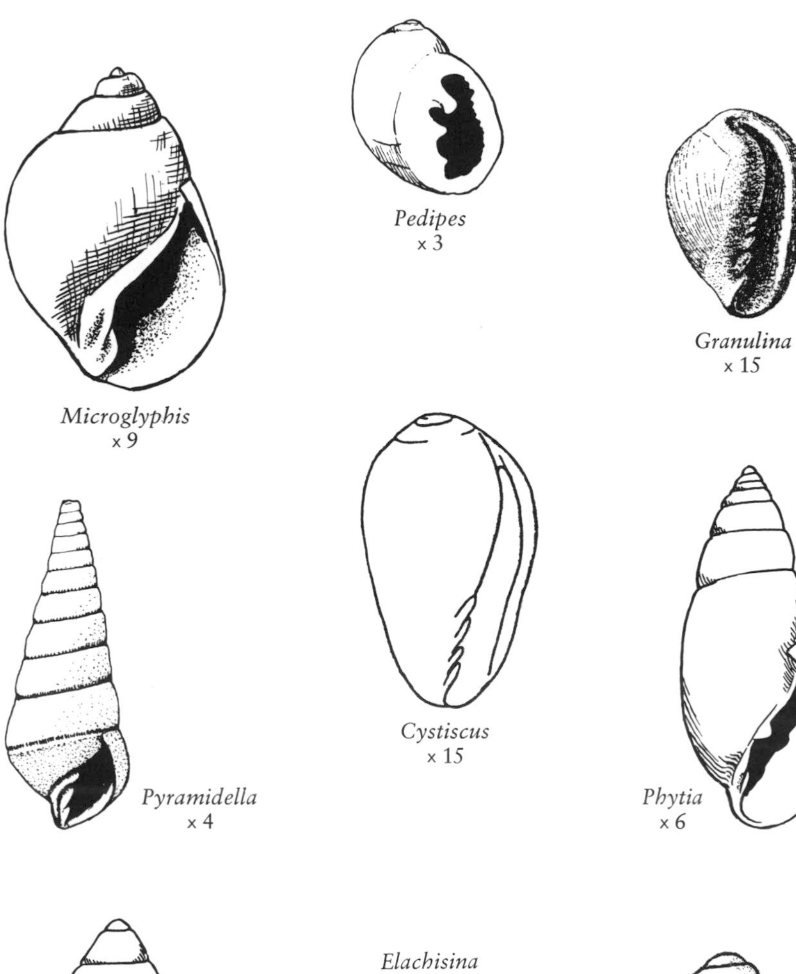

Pedipes
x 3

Granulina
x 15

Microglyphis
x 9

Cystiscus
x 15

Pyramidella
x 4

Phytia
x 6

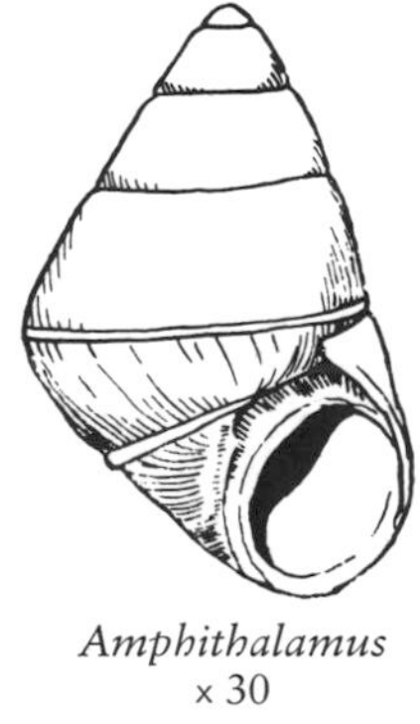

Amphithalamus
x 30

Elachisina
x 9

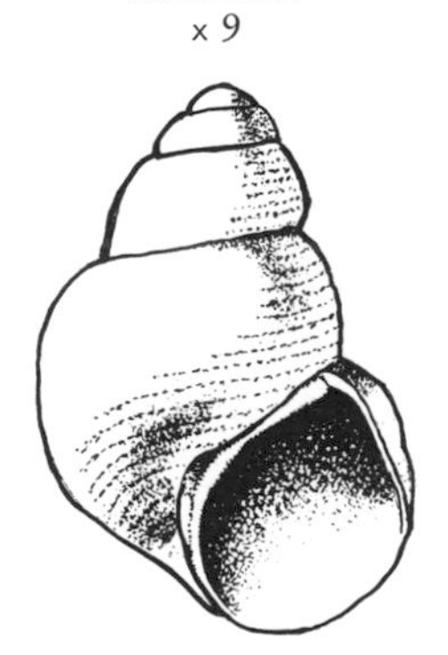

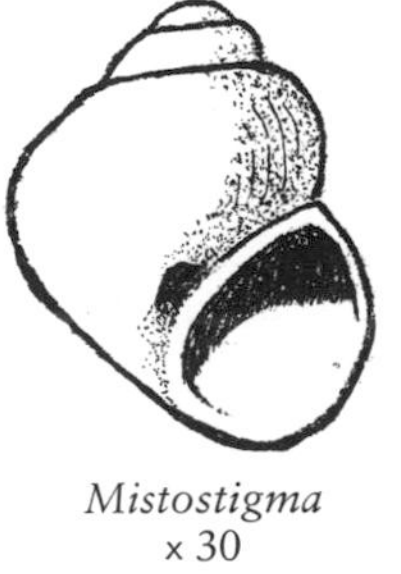

Mistostigma
x 30

115(111) Columellar folds two ... *Microglyphis*
Columellar folds three or more ... 116
116(115) Spire low, of few turns ... 117
Spire moderate to high, of several turns ... 119
117(116) With three large folds, uppermost heaviest ... *Pedipes*
With several small folds on lower part of columella ... 118
118(117) Spire concealed by body whorl ... *Granulina*
Spire evident though low ... *Cystiscus**
119(116) Folds near anterior end of columella; nuclear whorls heterostrophic (apparently sinistrally coiled) ... *Pyramidella*
Folds mainly above anterior end of columella; nuclear whorls normally coiled, dextral ... *Phytia*
120(108) Outline globose to turbinate, spire of few whorls ... 121
Outline slender-ovate, spire with several whorls (mostly more than six) ... 140
121(120) Adults small to minute (height less than 5 mm) ... 122
Adults medium-sized (height more than 5 mm) ... 129
122(121) Inner lip separated from columella by a shelf ... *Amphithalamus*
Inner lip continuous with columella ... 123
123(122) Umbilicate (at least with an umbilical chink) ... 124
Nonumbilicate ... 126
124(123) Surface with minutely punctate spiral lines ... *Elachisina*
Surface smooth or with other than punctate sculpture ... 125
125(124) Minute (height less than 2 mm), whorls few (less than three) ... *Mistostigma*
Small (height 2 to 4 mm), whorls four or more ... *Rissoella*
126(123) With prominent median carina ... *Anabathron*
Whorls rounded, not carinate ... 127
127(126) Inner lip spread out as a callus ... *Assiminea*
Inner lip not callused ... 128

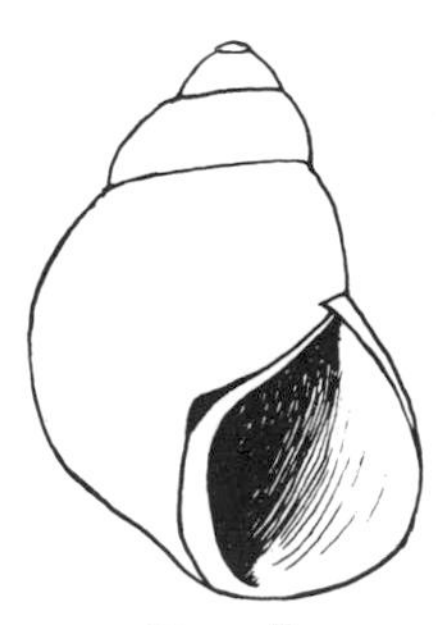

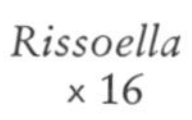

Rissoella
x 16

Anabathron
x 45

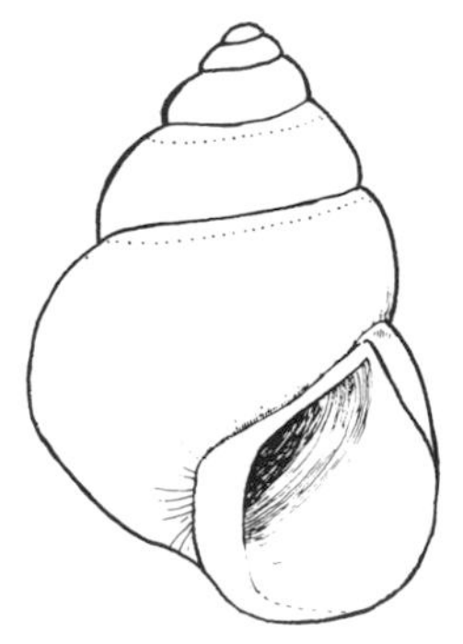

Assiminea
x 15

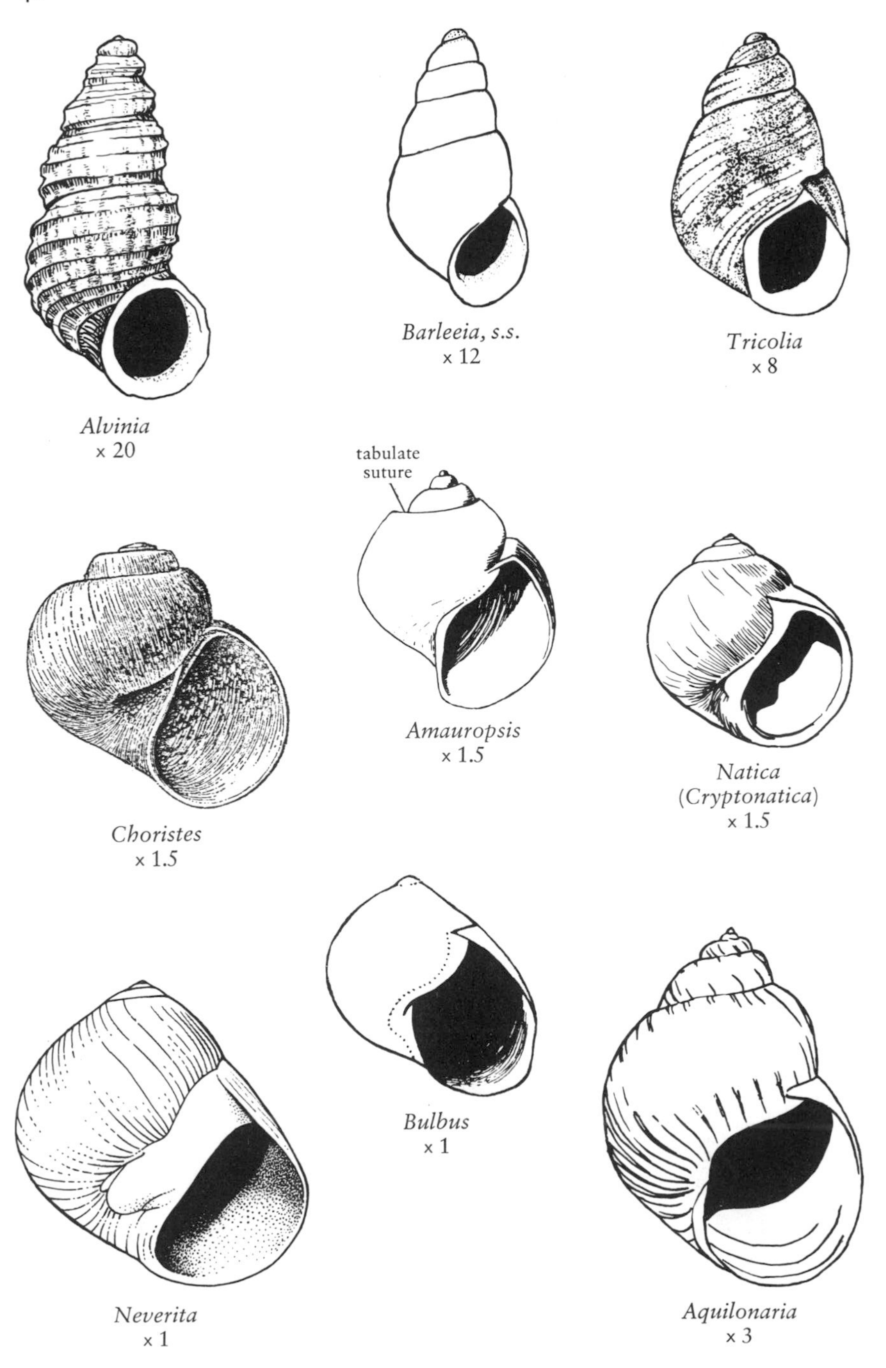

Alvinia
x 20

Barleeia, s.s.
x 12

Tricolia
x 8

Choristes
x 1.5

Amauropsis
x 1.5

Natica
(*Cryptonatica*)
x 1.5

Neverita
x 1

Bulbus
x 1

Aquilonaria
x 3

128(127) Shell strongly sculptured, nucleus smooth or with spiral ribbing . . . . *Alvinia*
Shell smooth, nuclear whorls pitted . . . . *Barleeia, s.s.*
129(121) Color pattern of irregular spots or stripes (mostly reddish) . . . . *Tricolia*
Color pattern, if present, not irregular 130
130(129) Suture channeled or tabulate 131
Suture not channeled nor tabulate 132
131(130) Base widely umbilicate . . . . *Choristes*
Base narrowly umbilicate or entire . . . . *Amauropsis*
132(130) Umbilical area covered by callus or showing at most only a chink at inner lip margin 133
Shell clearly umbilicate 139
133(132) With callus pad covering umbilical area 134
With a narrow chink or nonumbilicate 135
134(133) Callus plug at lower edge of inner lip; operculum exteriorly calcareous . . . . *Natica (Cryptonatica)*
Callus extending from upper edge of inner lip; operculum horny . . . . *Neverita*
135(133) With conspicuous periostracum 136
Without visible periostracum 137
136(135) Whorls few, periostracum smooth . . . . *Bulbus*
Whorls several, periostracum rough, shaggy . . . . *Aquilonaria*
137(135) Inner lip smooth, appressed . . . . *Littorina, s.l.*
Inner lip with narrow chink 138
138(137) Inner lip with simple chink separating it from body whorl . . . . *Littorina (Algamorda)*
Inner lip grooved, guttered, or well separated from base of body whorl . . . . *Lacuna*

Littorina, s.l.
× 2.5

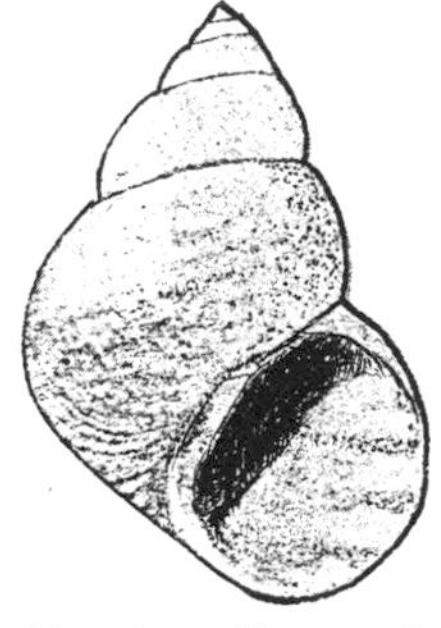

Littorina (*Algamorda*)
× 6

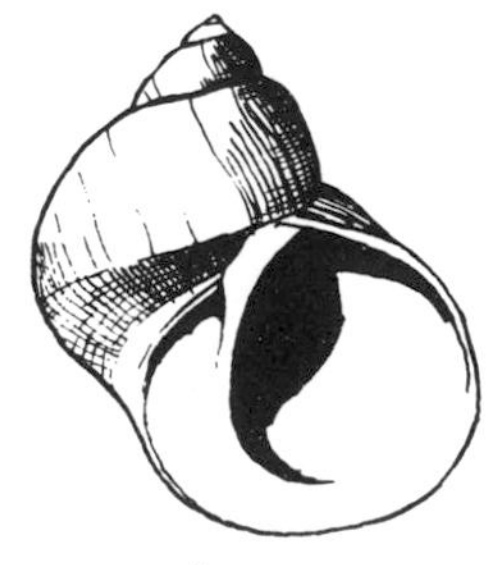

Lacuna
× 6

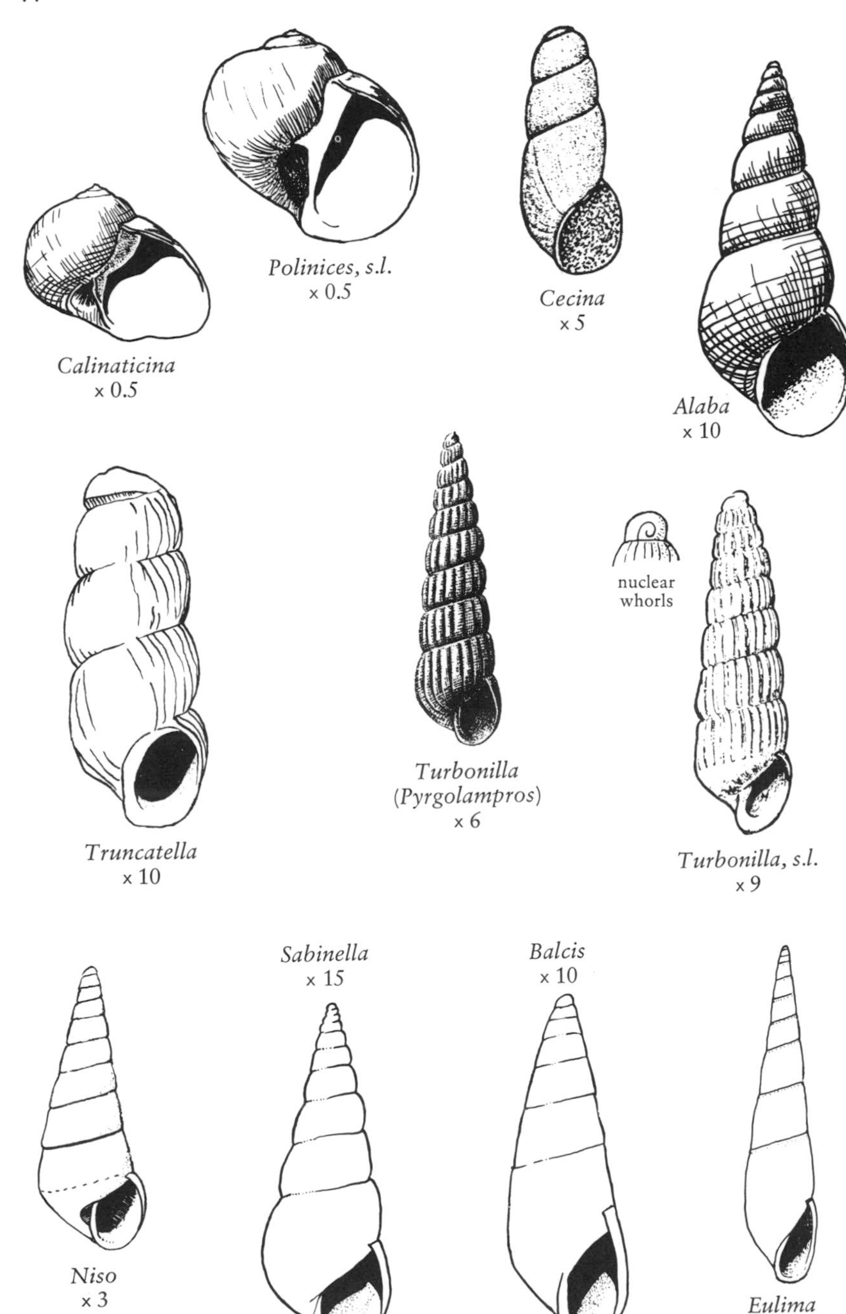

Calinaticina × 0.5

Polinices, s.l. × 0.5

Cecina × 5

Alaba × 10

Truncatella × 10

Turbonilla (Pyrgolampros) × 6

Turbonilla, s.l. × 9

Niso × 3

Sabinella × 15

Balcis × 10

Eulima × 7

139(132) Inner lip thin . . . . *Calinaticina*
Inner lip with callus partly covering upper edge of umbilicus . . .
. . . . *Polinices, s.l.*
140(120) Spire tip truncated (broken off and discarded) in adult 141
Spire tip persistent in adult 142
141(140) Whorls smooth throughout . . . . *Cecina*
Whorls with weak to strong axial ribs . . . . *Truncatella*
142(140) Nuclear whorls heterostrophic . . . . *Turbonilla*
Nuclear whorls not heterostrophic 143
143(142) Surface of shell smooth or nearly so 144
Surface of shell with well-developed sculpture on some part 152
144(143) Surface glassy-white, polished 145
Surface not glassy-white, not highly polished 149
145(144) With a few irregular axial swellings on spire and fine spiral striae on body whorl . . . . *Alaba**
Without axial swellings or spiral striae 146
146(145) Shell umbilicate . . . . *Niso*
Shell nonumbilicate 147
147(146) Whorls inflated . . . . *Sabinella*
Whorls flattened, sutures little impressed 148
148(147) Shell only moderately slender, aperture oval . . . . *Balcis*
Shell very slender, aperture elongate-oval . . . . *Eulima*
149(144) Spire whorls flattened, periphery of body whorl somewhat angulate . . . . *Barleeia (Pseudodiala)*
Spire whorls inflated, periphery rounded 150
150(149) Peristome interrupted; surface with irregular wrinkles or weak sculpture below suture . . . . *Aclis, s.s.*
Peristome entire; surface uniform, either smooth or finely striate
. . . . 151
151(150) Minute (height less than 2 mm); spire cylindrical . . . .
. . . . *Nannoteretispira*
Small (height 2 mm or more); spire evenly conic . . . . *Cingula**

Barleeia (Pseudodiala) x 8

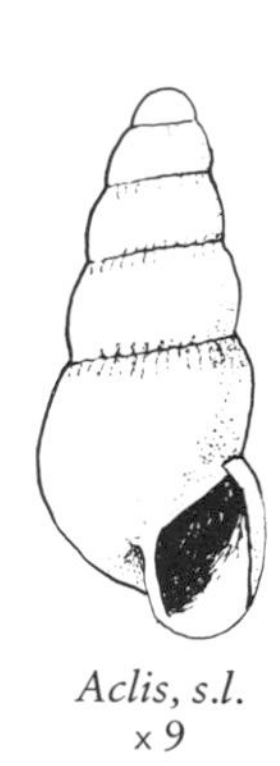
Aclis, s.l. x 9

Nannoteretispira x 21

Cingula x 10

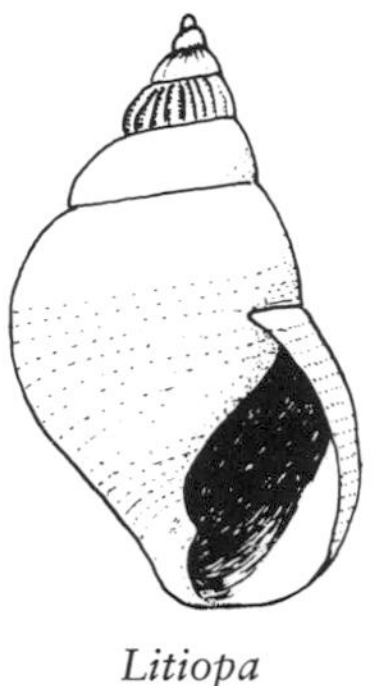

Litiopa
x 7

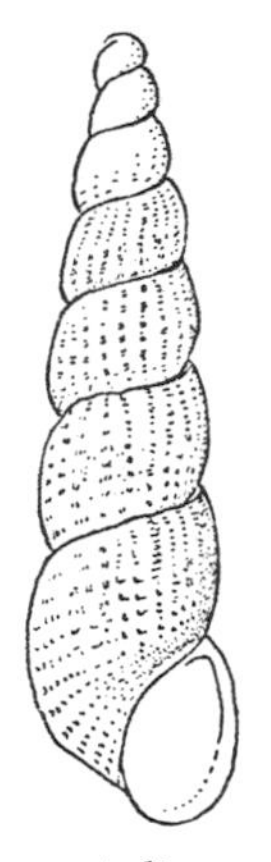

Aclis
(*Graphis*)
x 20

Rissoina
x 13

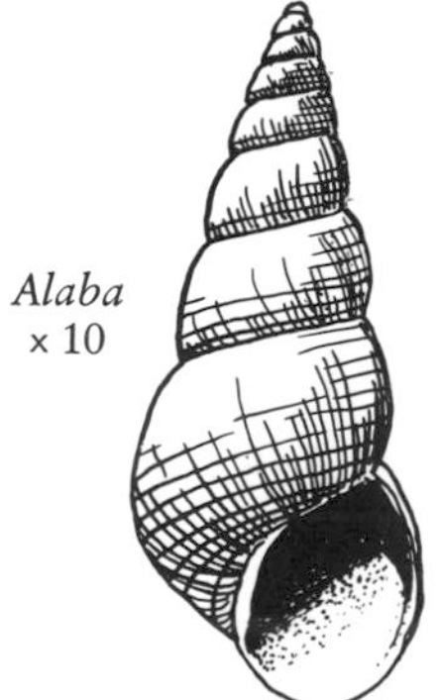

Alaba
x 10

Epitonium
x 2

Opalia
x 3

Turritella
x 1

Turritellopsis
x 7

Aclis
(*Schwengelia*)
x 12.5

152(143) Sculpture predominantly axial, at least on spire 153
Sculpture spiral or evenly reticulate throughout 158

153(152) Columella sharply cut off anteriorly . . . . *Litiopa*
Columella smoothly rounded, entire 154

154(153) Axial ribs vertically directed . . . . *Aclis (Graphis)*
Axial ribs slanting, not vertical 155

155(154) Axial ribs parallel to left side of shell; aperture half-moon-shaped . . . . *Rissoina*
Axial ribs parallel to right side of shell; aperture rounded 156

156(155) Axial sculpture of irregular low swellings or varices; outer lip thin . . . . *Alaba**
Axial sculpture of regular lamellae or costae; outer lip thickened 157

157(156) Axial costae thin, lamellar . . . . *Epitonium*
Axial costae rounded or subdued, not thin . . . . *Opalia*

158(152) Whorls more than ten 159
Whorls fewer than ten 160

159(158) Adult more than 25 mm in length . . . . *Turritella*
Adult less than 25 mm in length . . . . *Turritellopsis*

160(158) Each whorl smooth above a carina . . . . *Aclis (Schwengelia)*
Whorls with carina but lacking smooth band 161

161(160) Adults more than 10 mm in length . . . . *Tachyrynchus*
Adults less than 10 mm in length 162

162(161) Shell relatively sturdy, lip thick . . . . *Halistylus*
Shell relatively thin and delicate, lip thin 163

163(162) Short-ovate, with four to six whorls; inner lip rimate . . . . *Cingula**
Tapering, with seven to ten whorls; lip appressed . . . . *Alabina*

164(43) Length of aperture equal to longest dimension of shell; coiling involute 165
Length of aperture less than total length; coiling helical 168

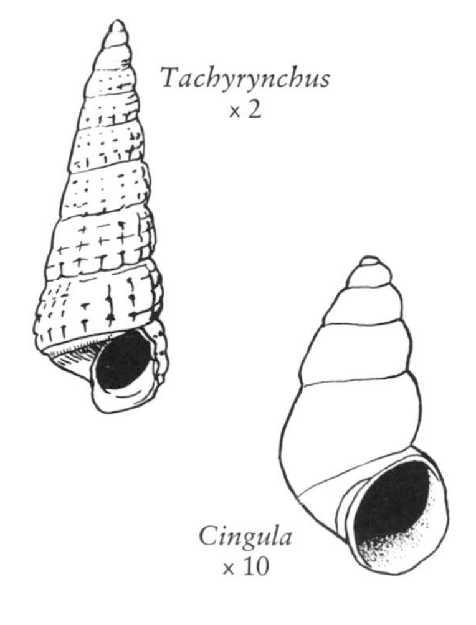
Tachyrynchus ×2

Cingula ×10

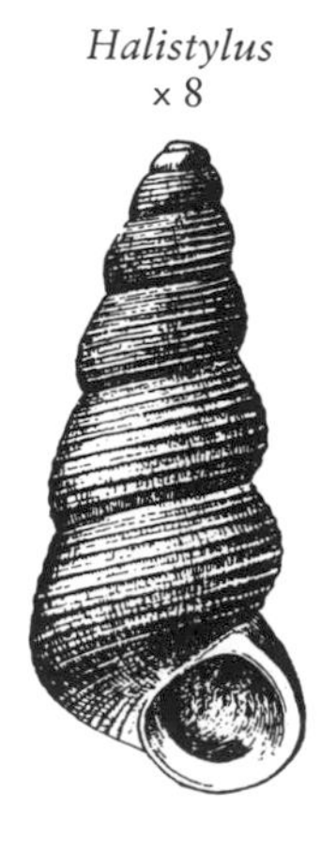
Halistylus ×8

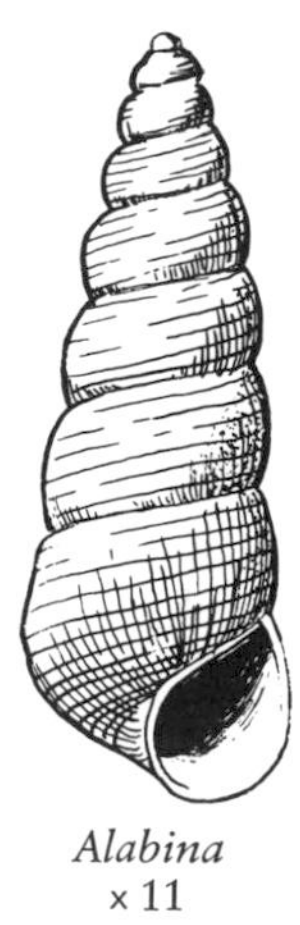
Alabina ×11

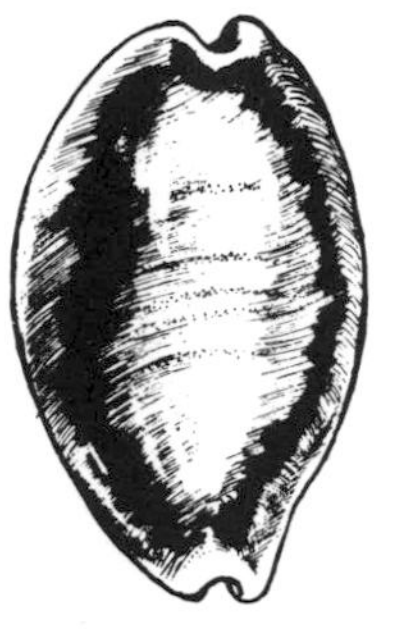

Cypraea (*Zonaria*)
× 0.8

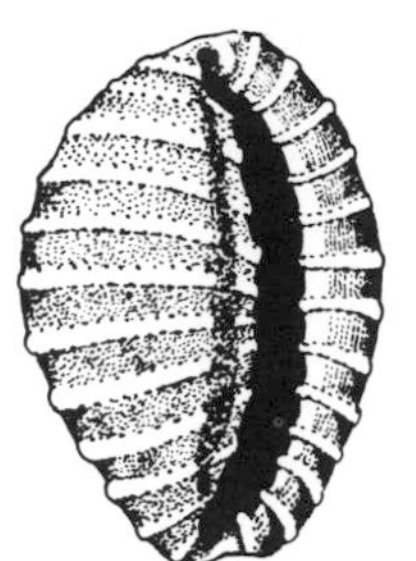

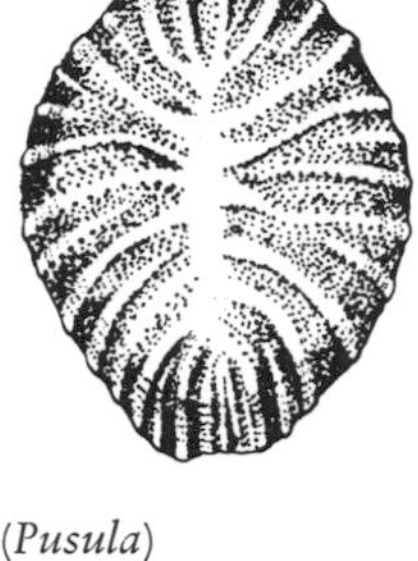

Trivia (*Pusula*)
× 4

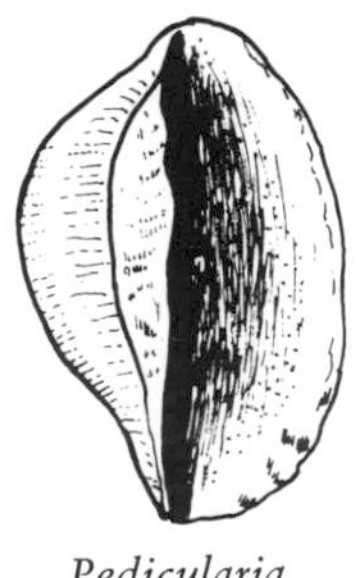

Pedicularia
× 3.5

Simnia
× 2

Volutharpa
× 1.5

Erato
× 8

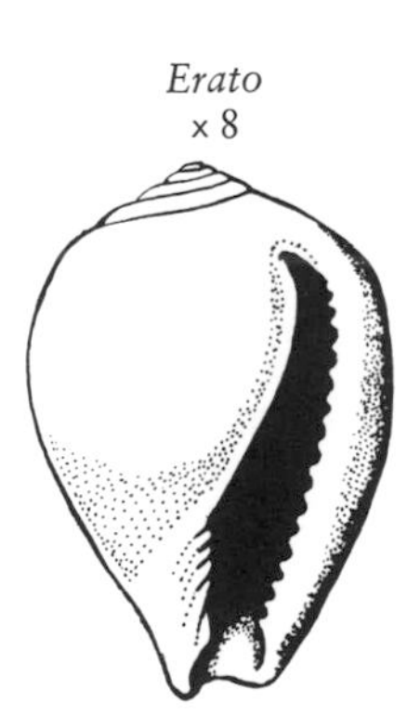

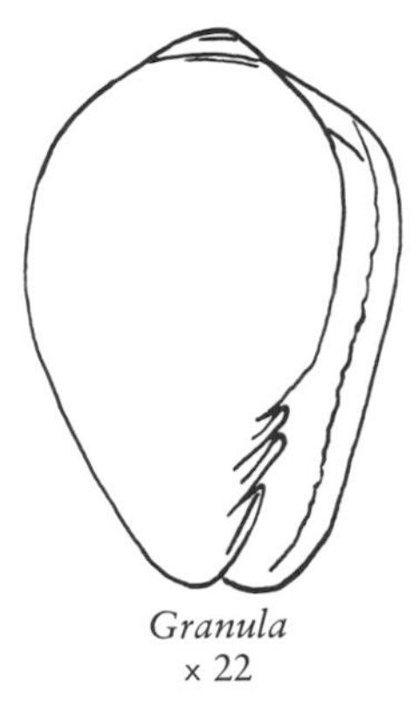

Granula
× 22

Terebra
× 1

165(164) Aperture slotlike, both lips denticulate 166
Aperture fairly wide, lips not denticulate 167
166(165) Shell smooth, shiny . . . . *Cypraea (Zonaria)*
Shell ribbed, back frequently with a longitudinal furrow . . . . *Trivia (Pusula)*
167(165) Minutely spirally striate; inner lip smooth . . . . *Pedicularia*
Smooth, not striate; inner lip twisted at posterior end . . . . . . *Simnia*
168(164) Aperture without a posterior siphonal notch 169
Aperture with posterior siphonal notch or slit at or near suture 236
169(168) Aperture relatively long, three-fourths or more length of shell. . . 170
Aperture less than three-fourths length of shell 172
170(169) Shell large, with fibrous periostracum . . . . *Volutharpa*
Shell medium to small, without periostracum 171
171(170) Shell pinkish-brown; length of adults 5 mm or more; outer lip markedly thickened . . . . *Erato*
Shell white to pellucid; length less than 5 mm; outer lip not markedly thickened . . . . *Granula*
172(169) Turriform, spire with numerous whorls 173
Not turriform, spire whorls not especially numerous 181
173(172) With columellar folds . . . . *Terebra*
Without columellar folds 174
174(173) With varices in the form of irregularly spaced axial swellings . . . . *Cerithidea*
Without axial swellings or varices 175
175(174) Canal short, poorly differentiated 176
Canal well developed, moderate to long 177
176(175) Aperture narrowed anteriorly . . . . *Bittium*
Aperture flared anteriorly . . . . *Diastoma*
177(175) Coiling sinistral . . . . *Triphora*
Coiling dextral 178

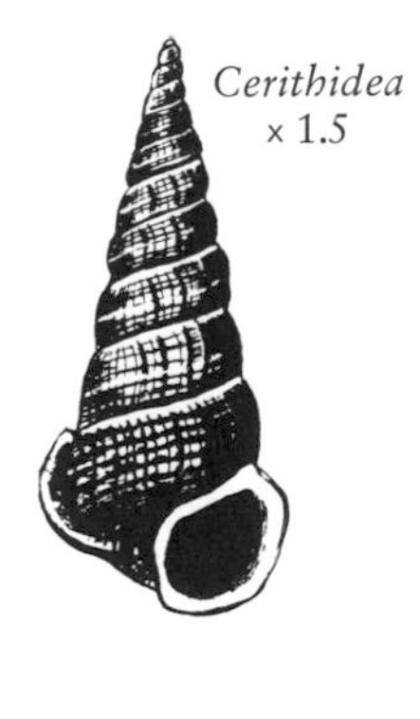
Cerithidea
× 1.5

Diastoma
× 7

Bittium
× 7

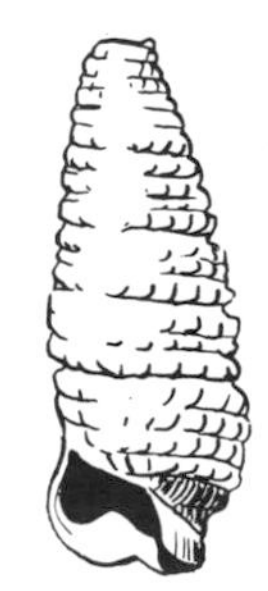
Triphora
× 8

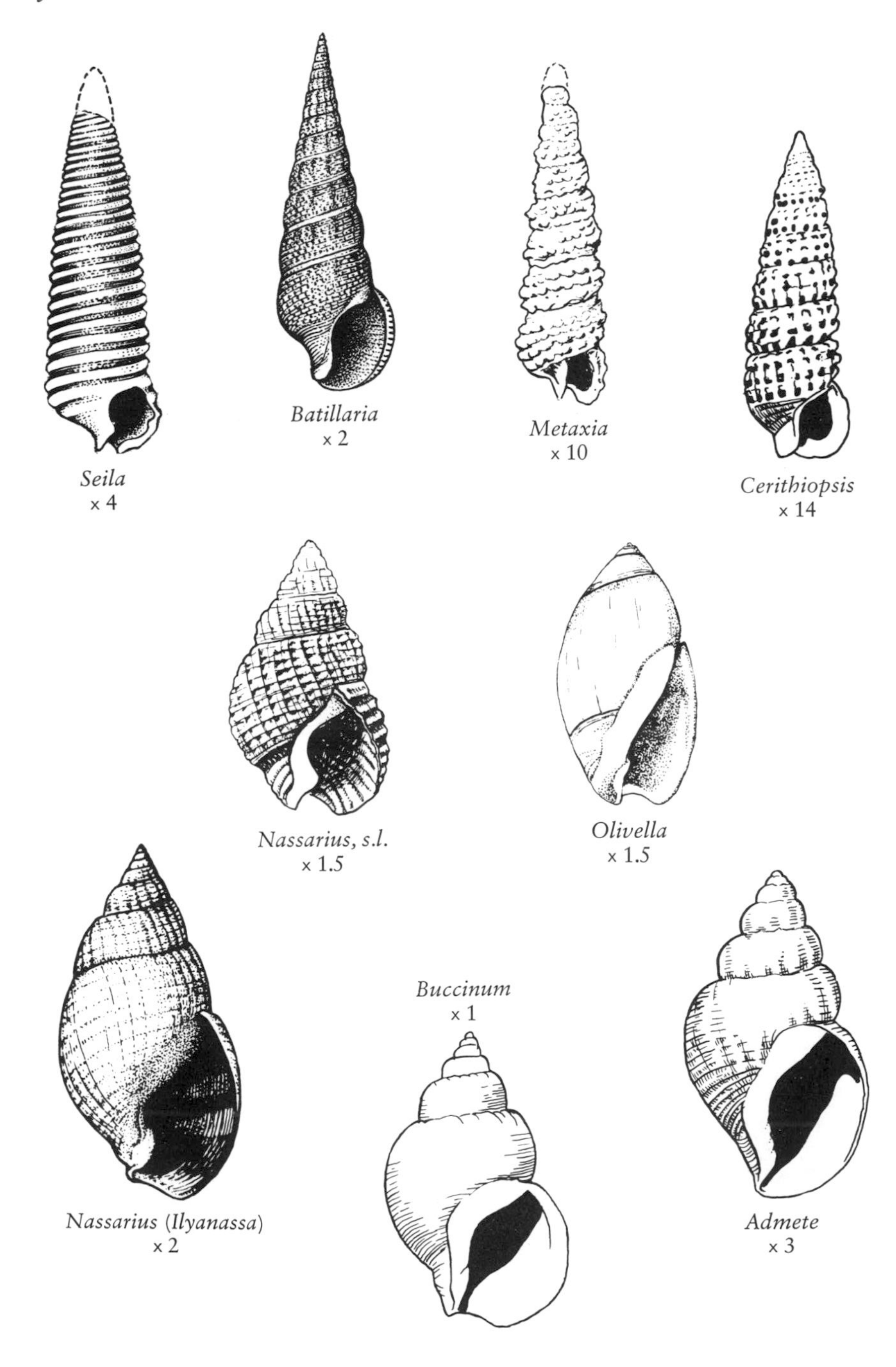
Seila
× 4
Batillaria
× 2
Metaxia
× 10
Cerithiopsis
× 14
Nassarius, s.l.
× 1.5
Olivella
× 1.5
Nassarius (Ilyanassa)
× 2
Buccinum
× 1
Admete
× 3

178(177) Whorls flattened; sculpture entirely spiral . . . . *Seila*
Whorls somewhat inflated; sculpture both axial and spiral 179
179(178) Spire with axial ribs that fade out on lower whorls . . . . . . *Batillaria*
Sculpture cancellate throughout 180
180(179) Suture deeply channeled; axial ribs weak, spiral ribs with elongate nodes . . . . *Metaxia*
Suture only moderately impressed; axial and spiral ribs of nearly equal strength, nodes rounded . . . . *Cerithiopsis*
181(172) Anterior canal set off from body whorl by a constriction or furrow (a fossa) . . . . *Nassarius, s.l.*
Anterior canal not set off by a fossa 182
182(181) Columella with one or more folds 183
Columella smooth except for occasional denticles 193
183(182) Columella with a single twisted plait at the anterior end 184
Columella with two or more folds 186
184(183) Shell smooth, highly polished . . . . *Olivella*
Shell with sculpture evident, not polished 185
185(184) Aperture blackish-glazed within . . . . *Nassarius (Ilyanassa)*
Aperture light-colored within, not glazed . . . . *Buccinum*
186(183) Outer lip smooth within or merely reflecting exterior sculpture 187
Outer lip lirate within 191
187(186) Canal indistinct, columella obliquely truncate . . . . *Admete*
Canal distinct, no columellar truncation 188
188(187) Surface of shell smooth . . . . *Arctomelon*
Surface sculptured 189
189(188) With strong axial ribs in addition to spiral ribs . . . . *Metzgeria*
Sculpture predominantly spiral 190

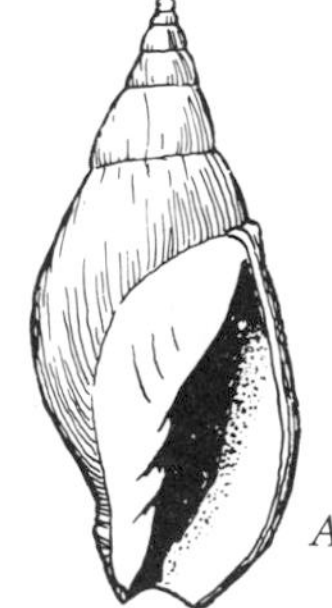

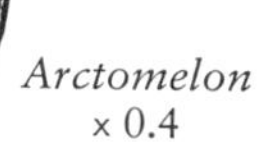

Arctomelon
x 0.4

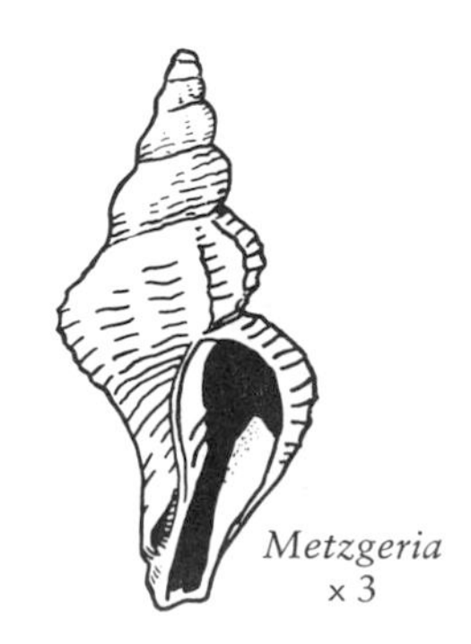

Metzgeria
x 3

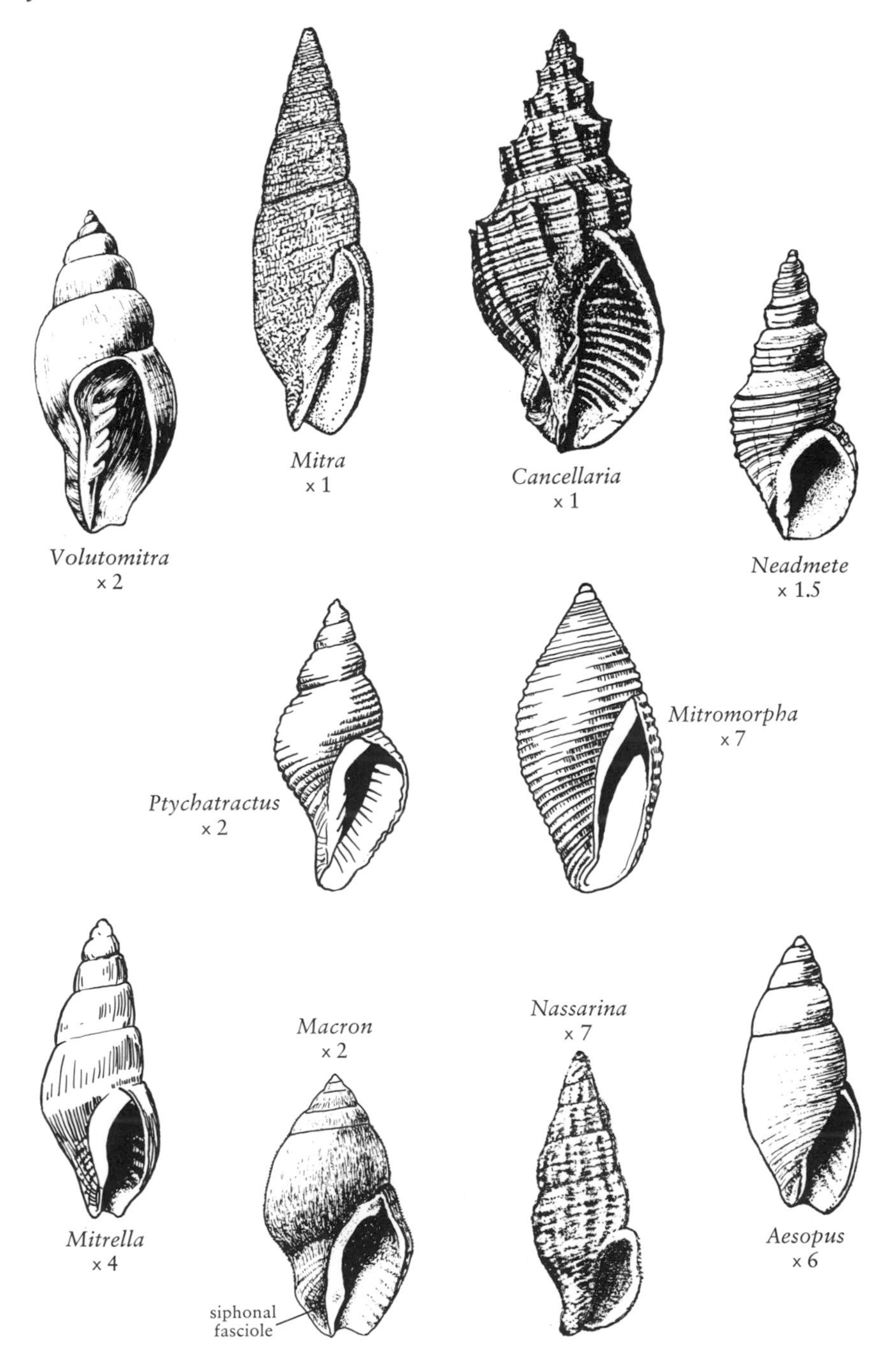
Volutomitra
× 2
Mitra
× 1
Cancellaria
× 1
Neadmete
× 1.5
Ptychatractus
× 2
Mitromorpha
× 7
Mitrella
× 4
Macron
× 2
siphonal
fasciole
Nassarina
× 7
Aesopus
× 6

190(189) Striae minute; columellar folds of equal strength ..... *Volutomitra*
Striae moderate to coarse; posterior columellar fold the strongest .... *Mitra*

191(186) Columellar folds well developed .... *Cancellaria*
Columellar folds weak 192

192(191) Shell slender, aperture less than one-half length .... *Neadmete*
Shell biconic, aperture one-half length .... *Ptychatractus*

193(182) Anterior canal short to almost obsolete, outer lip rounding smoothly to anterior end of shell 194
Anterior canal moderate to long, outer lip sinuous 212

194(193) Aperture elongate, narrow, more than one-half length of shell .... *Mitromorpha*
Aperture ovate to quadrate, one-half the length of the shell or less 195

195(194) Base with spiral ribs, spire smooth 196
Base not more strongly sculptured than remainder of the shell 197

196(195) Siphonal fasciole well developed; shell length 20 mm or more; whorls inflated .... *Macron*
Siphonal fasciole wanting; shell length less than 15 mm; whorls mostly flat-sided .... *Mitrella*

197(195) Aperture short (about one-third length of shell); shell less than 10 mm in length 198
Aperture more than one-third length of shell; shell length more than 10 mm 199

198(197) Sculpture strongly developed .... *Nassarina (Zanassarina)*
Sculpture weak to wanting .... *Aesopus*

199(197) Anterior canal spoutlike 200
Anterior canal slotlike or broad, not spoutlike 201

200(199) Spiral ribs with scaly growth-lamellae; no periostracum present .... *Latiaxis*
Spiral ribs without scaly lamellae; a fibrous periostracum present in most forms .... *Trichotropis*

Latiaxis
× 1.5

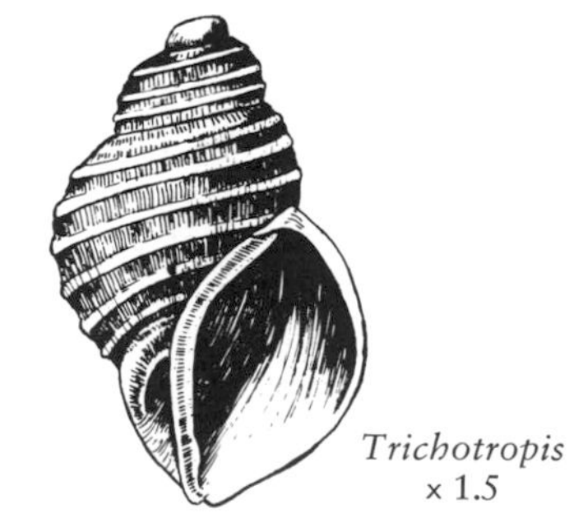

Trichotropis
× 1.5

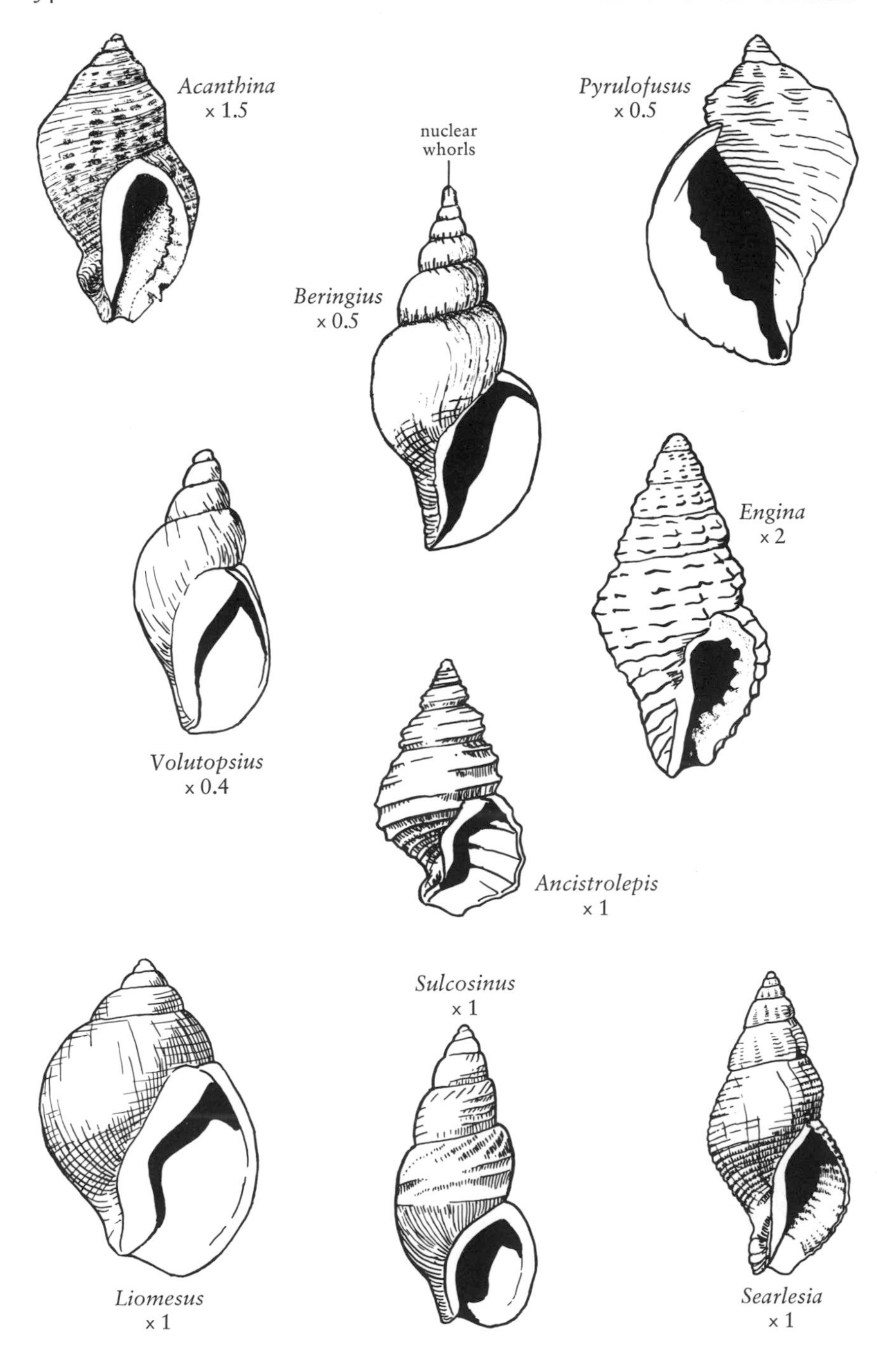
Acanthina
x 1.5
Pyrulofusus
x 0.5
nuclear
whorls
Beringius
x 0.5
Engina
x 2
Volutopsius
x 0.4
Ancistrolepis
x 1
Sulcosinus
x 1
Liomesus
x 1
Searlesia
x 1

201(199) Outer lip with a single strong tooth near canal in adults...... .... *Acanthina*
Outer lip smooth or evenly denticulate 202
202(201) Coiling sinistral .... *Pyrulofusus*
Coiling dextral 203
203(202) Nuclear whorls large, either cylindrical or bulbous 204
Nuclear whorls not conspicuous 205
204(203) Aperture long, more than one-half shell length .... *Volutopsius*
Aperture not more than one-half shell length .... *Beringius*
205(203) Outline biconic; denticles present in posterior part of aperture .... *Engina*
Outline not biconic; lips smooth or denticles, if present, not in posterior part of aperture 206
206(205) Anterior canal markedly broad and open 207
Anterior canal narrow or slotlike 209
207(206) Shell with spiral ribs; canal twisted to left .... *Ancistrolepis*
Shell smooth or nearly so; canal straight 208
208(207) Spire elevated, slender; aperture short .... *Sulcosinus*
Spire short; shell and aperture long and broad .... *Liomesus*
209(206) Spire and base differently sculptured 210
Spire and base with similar sculpture 211
210(209) Body whorl with spiral sculpture only .... *Searlesia*
Body whorl with both spiral and axial sculpture above periphery, spiral only below .... *Amphissa*
211(209) Small; axial ribs nodosely sculptured .... *Trachypollia*
Medium-sized to large; sculpture spiral or with scaly axial lamellae, not nodose .... *Nucella*
212(193) With varices (resting stages) at regular intervals on spire and body whorl 213
Axial sculpture consisting of ribs or simple lamellae, not heavy thickenings 221

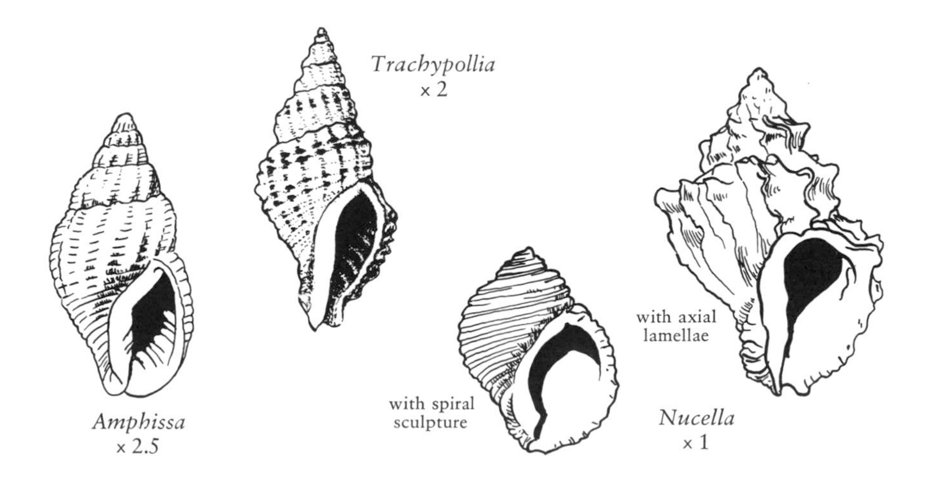

Trachypollia x 2

Amphissa x 2.5

Nucella x 1

Cymatium
× 0.5

Fusitriton
× 0.5

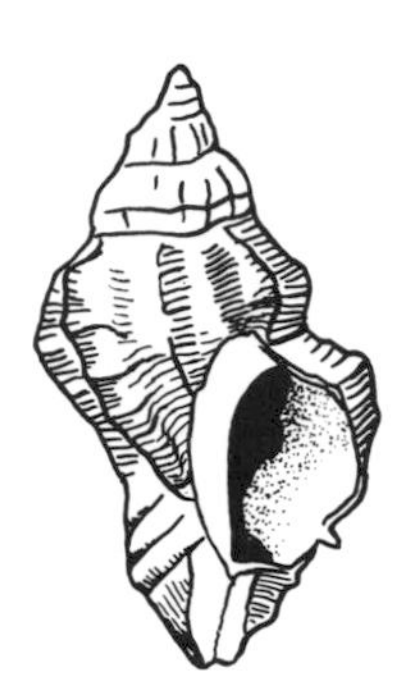
Ceratostoma
× 1

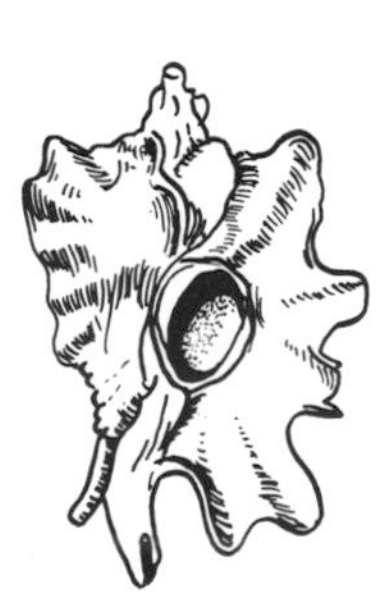
Pteropurpura, s.s.
× 0.7

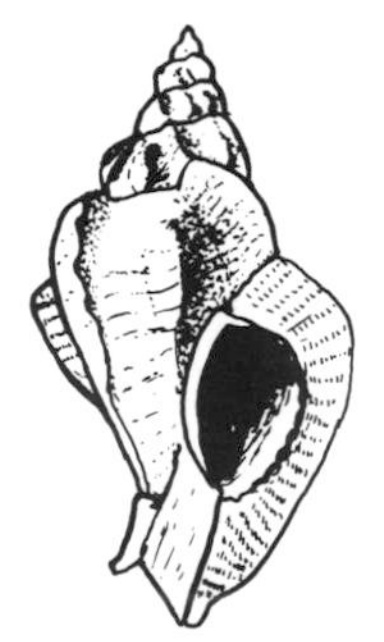
Pteropurpura
(*Shaskyus*)
× 1

Pterynotus
(*Pterochelus*)
× 1

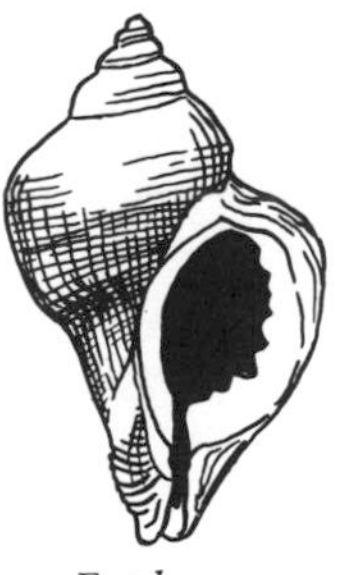
Eupleura
× 1.5

Murexiella
× 1

213(212) With hairy periostracum 214
Without hairy periostracum 215
214(213) Outer lip internally dentate . . . . *Cymatium*
Outer lip internally smooth . . . . *Fusitriton*
215(213) Varices three in number 216
Varices more or fewer than three 219
216(215) With a tooth on outer lip . . . . *Ceratostoma*
Without a conspicuous tooth on outer lip 217
217(216) Varices lamellar or winglike . . . . *Pteropurpura, s.s.*
Varices not lamellar or winglike 218
218(217) Varices recurved, low . . . . *Pteropurpura (Shaskyus)*
Varices with an open spine at shoulder . . . *Pterynotus (Pterochelus)*
219(215) Varices fewer than three, not buttressed . . . . *Eupleura*
Varices more than three, buttressed 220
220(219) Varix face intricately sculptured . . . . *Murexiella*
Varix face smooth or nearly so . . . . *Maxwellia*
221(212) Spire with eight or more whorls . . . . *Exilioidea*
Spire with fewer than eight whorls 222
222(221) Outer lip dentate or lirate within 223
Outer lip smooth or merely reflecting surface sculpture 227
223(222) Siphonal fasciole strong 224
Siphonal fasciole weak or wanting 226
224(223) Suture normally incised; sculpture scaly, often reticulate . . . . *Ocenebra*
Suture sinuous, apparently riding up on previous whorls; sculpture not scaly nor reticulate 225

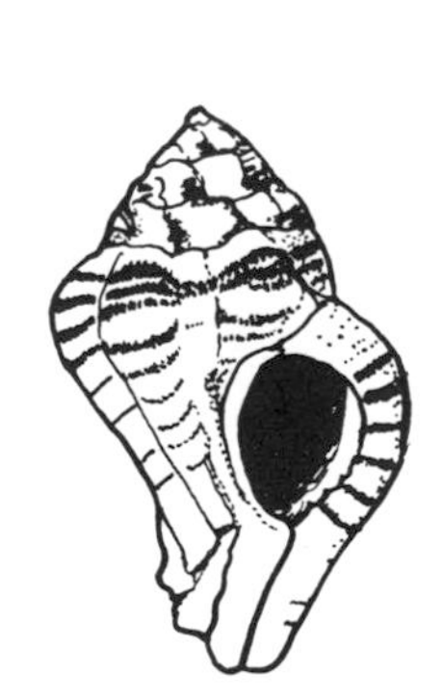
Maxwellia
×1

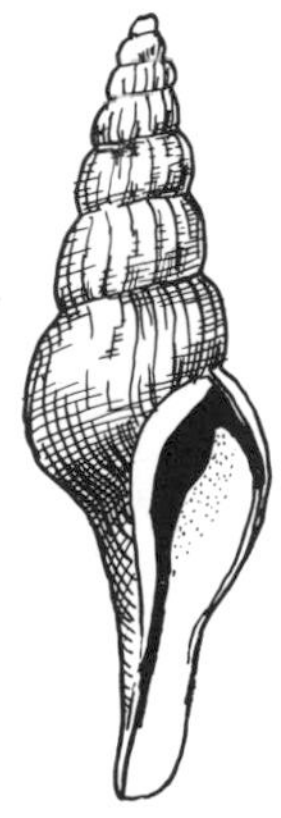
Exilioidea
×2

Ocenebra
×2

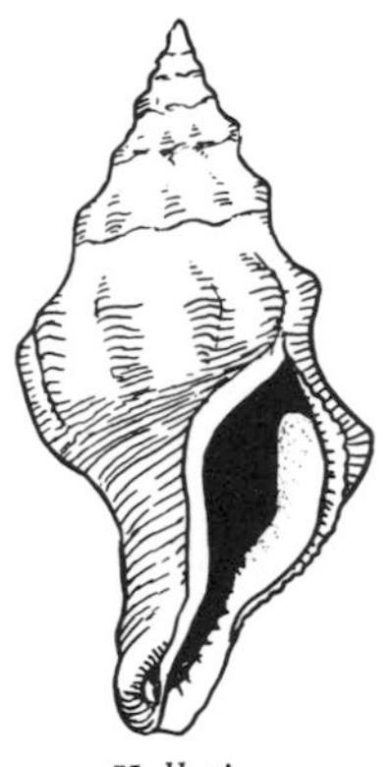

Kelletia
× 0.5

Roperia
× 1

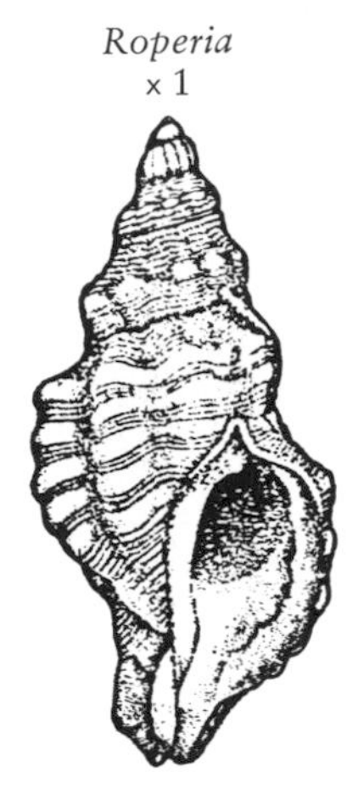

Fusinus
× 1.5

Urosalpinx
× 1.5

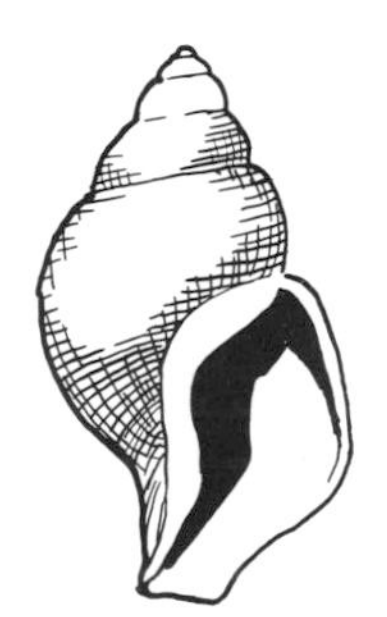

Colus
× 1

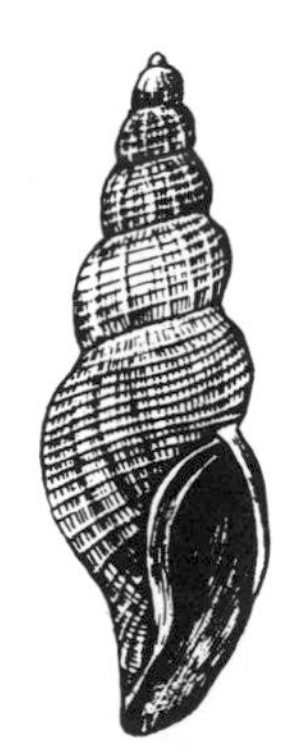

Morrisonella
× 2

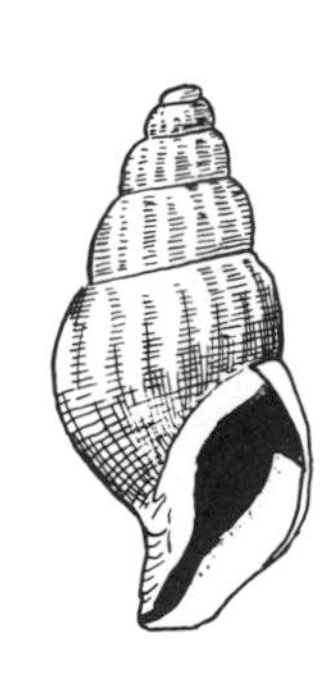

Mohnia
× 1

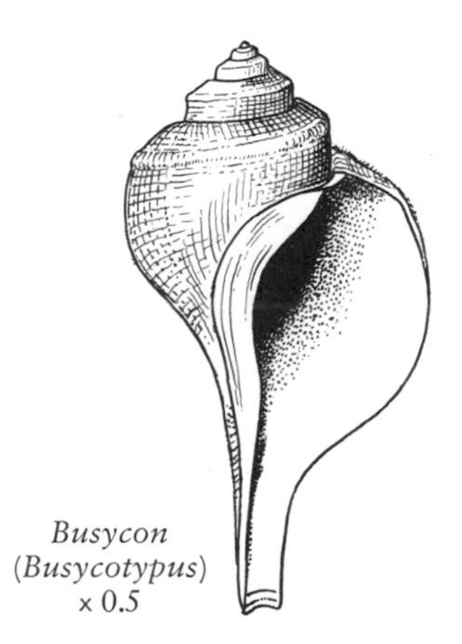

Busycon
(Busycotypus)
× 0.5

225(224) Shell large, heavy; outer lip smooth or lirate . . . . *Kelletia*
Shell medium-sized, not markedly heavy; outer lip denticulate within . . . . *Roperia*
226(223) Axial ribs well developed; canal long . . . . *Fusinus*
Axial ribs weak on body whorl; canal moderate in length . . . . *Urosalpinx*
227(222) With thin, adherent periostracum 228
Periostracum wanting 230
228(227) Spire smooth or with spiral sculpture only . . . . *Colus*
Spire with axial ribs 229
229(228) Slender, diameter about one-third length . . . . *Morrisonella*
Not slender, diameter about one-half length . . . . *Mohnia*
230(227) Sculpture entirely spiral 231
Sculpture not entirely spiral 232
231(230) Shell thin for its size; spire low . . . . *Busycon (Busycotypus)*
Shell sturdy; spire elevated . . . . *Neptunea*
232(230) Axial sculpture as low ridges 233
Axial sculpture as raised laminae 234
233(232) Canal moderately long, somewhat broad . . . . *Plicifusus*
Canal long and narrowed . . . . *Surculina*
234(232) Outer lip with strong tooth near anterior canal . . . . *Forreria*
Outer lip smooth, without teeth 235

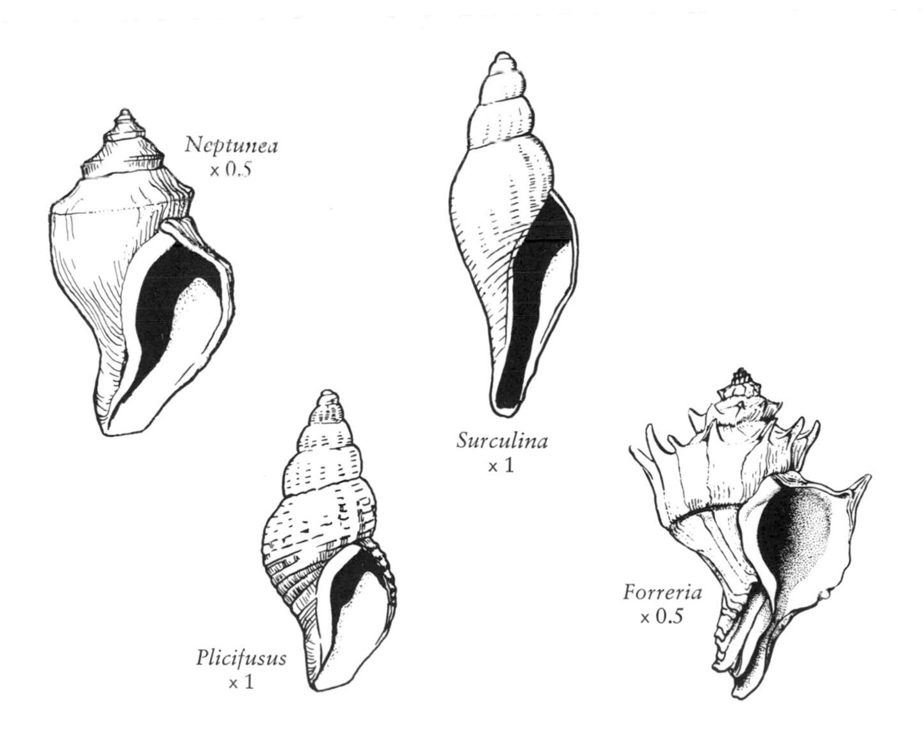

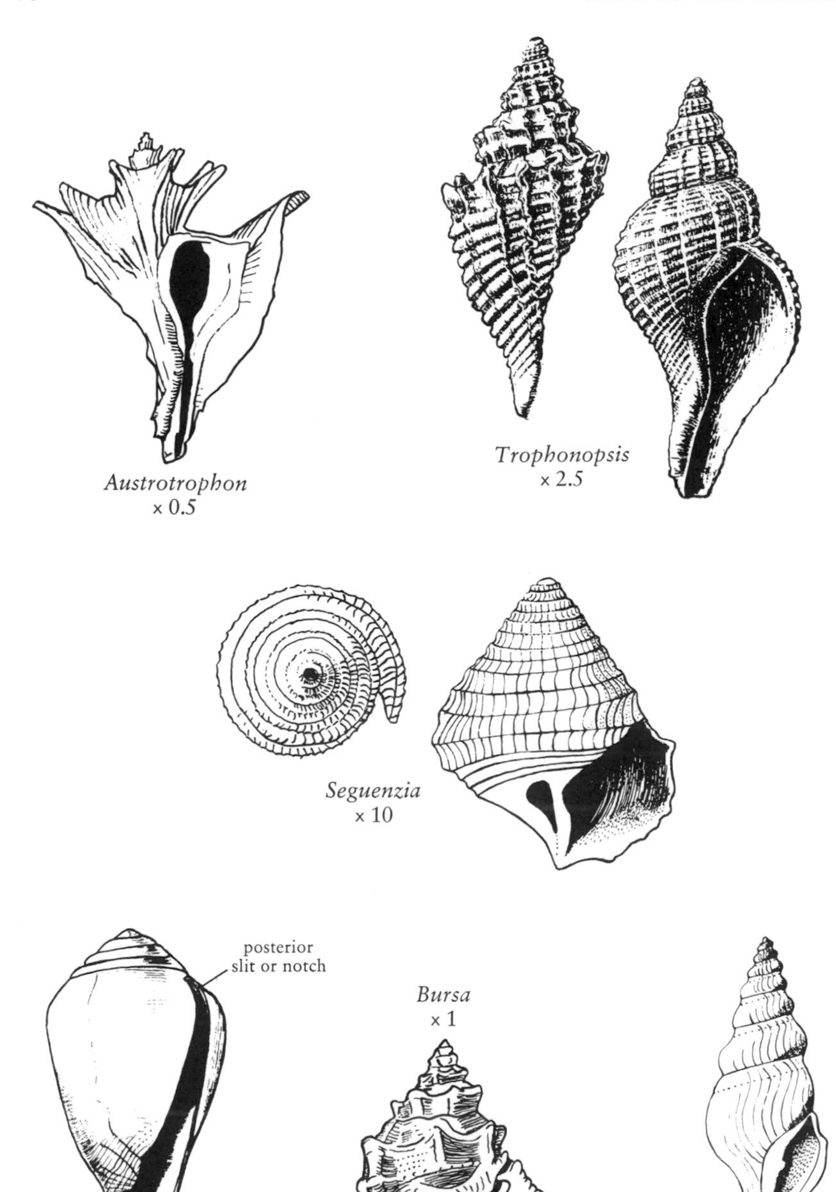

Austrotrophon
× 0.5

Trophonopsis
× 2.5

Seguenzia
× 10

Bursa
× 1

Conus
× 2

Borsonella
× 2

235(234) Axial lamellae guttered, widely flaring above aperture . . . . . . *Austrotrophon*
Axial lamellae rounded to slightly recurved, not widely flaring above aperture . . . . . . *Trophonopsis*
236(168) Shell obconic . . . . . . *Conus*
Shell not obconic 237
237(236) Shell globose to conic . . . . . . *Seguenzia*
Shell neither globose nor conic 238
238(237) With true varices (resting stages) . . . . . . *Bursa*
Without varices 239
239(238) Upper part of columella with a fold . . . . . . *Borsonella*
No fold on columella 240
240(239) Aperture at least half total height of shell 241
Aperture less than half total height 244
241(240) Sculpture of well-developed spiral ribs . . . . . . *Leucosyrinx, s.l.*
[Includes *Aforia, Irenosyrinx,* and *Steiraxis,* all from deep water]
Sculpture either subdued or reticulate 242
242(241) Sculpture subdued, adult shell more than 25 mm high . . . . . . *Megasurcula*
Sculpture reticulate, shell less than 25 mm high 243
243(242) Spire evenly reticulate, base with spiral ribs only . . . . . . *Daphnella*
Axial ribs slightly stronger than spiral ribs throughout . . . . . . *Oenopota, s.l.**
[Includes *Granotoma, Nodotoma, Propebela,* etc.]
244(240) Aperture about half length of spire (one-third length of shell) . . . 245
Aperture only slightly shorter than spire (about two-fifths length of shell) 250

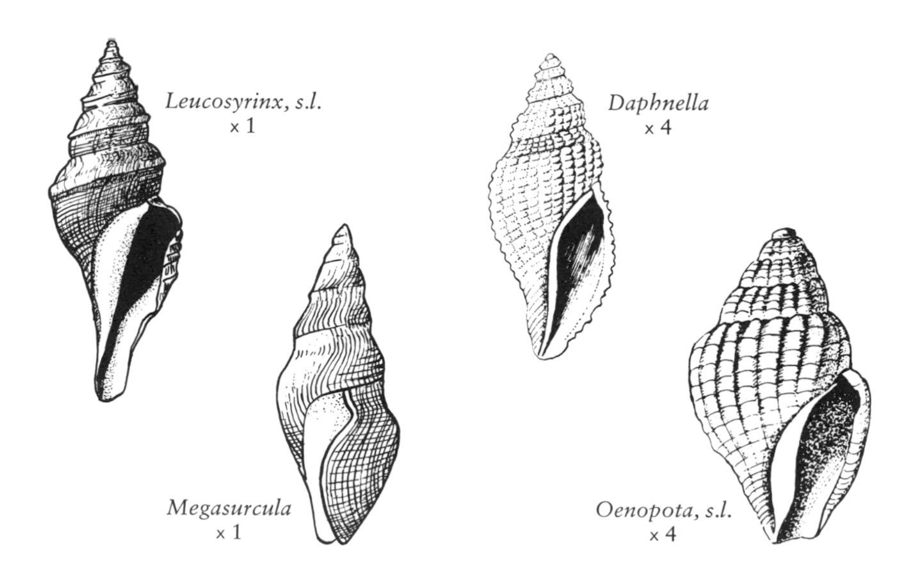

Antiplanes, s.s.
× 1.5

Suavodrillia
× 2

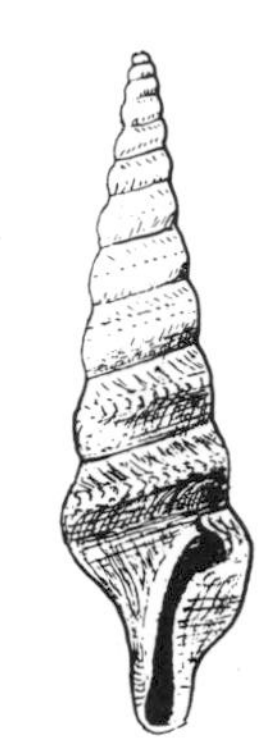

Antiplanes
(Rectiplanes)
× 1.5

Antiplanes
(Rectisulcus)
× 2

Elaeocyma
× 2

Rhodopetoma
× 2.5

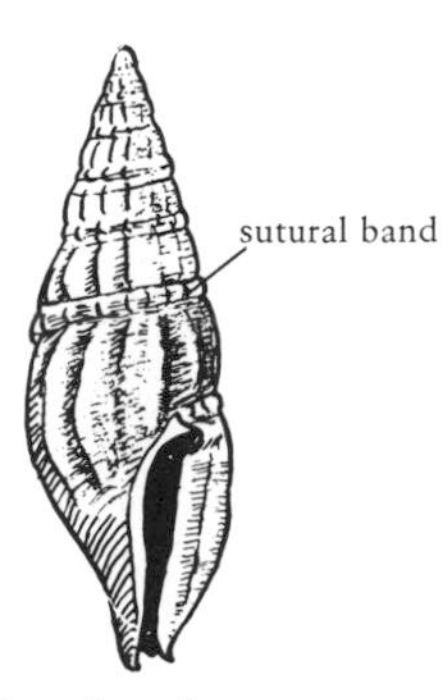

Pseudomelatoma
× 1.5

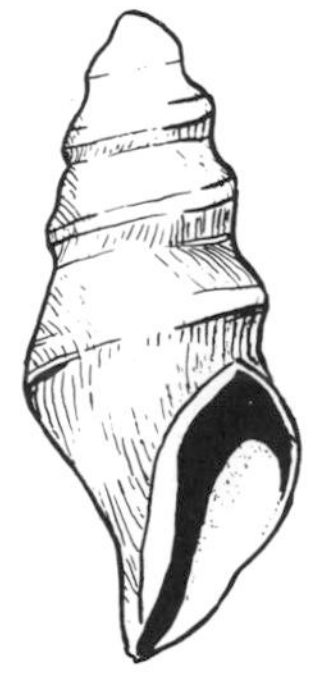

Carinoturris
× 2.5

245(244) Coiling sinistral . . . . *Antiplanes, s.s.*
Coiling dextral 246
246(245) Sculpture of growth lines only . . . . *Antiplanes (Rectiplanes)*
Sculpture stronger than growth lines only 247
247(246) Sculpture of spiral ribs only 248
Sculpture not entirely spiral 249
248(247) Whorls flat-sided; suture not indented . . . . *Suavodrillia*
Whorls convex in profile, suture indented . . . .
. . . . *Antiplanes (Rectisulcus)*
249(247) Surface with an oily gloss; axial ribs strong, oblique, especially on spire . . . . *Elaeocyma*
Shell covered with a periostracum; axial ribs faint, especially on anterior portions of whorls . . . . *Rhodopetoma*
250(244) Whorls carinate; posterior notch on a carina . . . . *Carinoturris*
Whorls not carinate 251
251(250) With a well-developed band below suture, resembling a second suture 252
No sutural band present 254
252(251) Sutural band beaded; parietal area smooth . . . . . . *Pseudomelatoma*
Sutural band not beaded; parietal area with callus 253
253(252) With a parietal tooth; outer lip entire . . . . *Crassispira, ss.*
Parietal callus smooth; outer lip with a secondary anterior notch . . . . *Crassispira (Burchia)*
254(251) Posterior notch deep; outer lip somewhat thickened 255
Posterior notch shallow to moderate; outer lip not thickened . . . 257
255(254) Protoconch whorls smooth, like nuclear tip . . . .
. . . . *Clathurella (Crockerella)*
Protoconch whorls carinate except for the smooth nuclear tip . . . 256

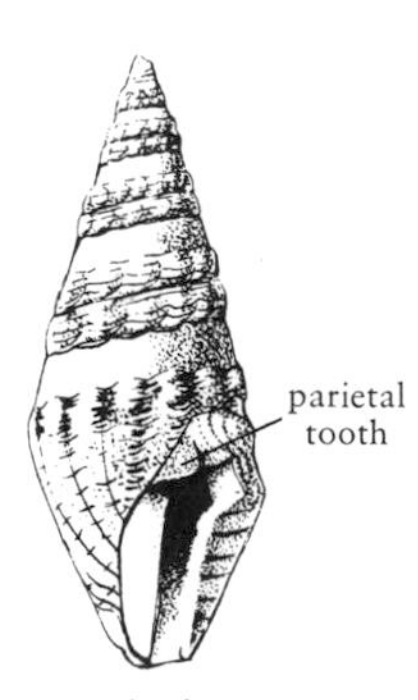

Crassispira, s.s.
× 2.5

Crassispira
(Burchia)
× 1

Clathurella
(Crockerella)
× 4

Cytharella
× 5

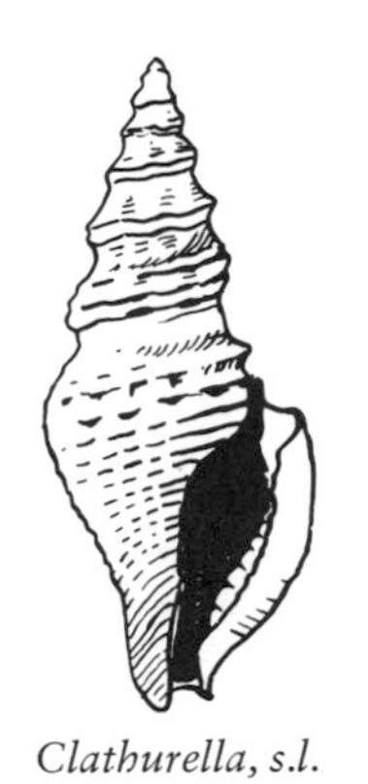

Clathurella, s.l.
× 4

Cymakra
× 4

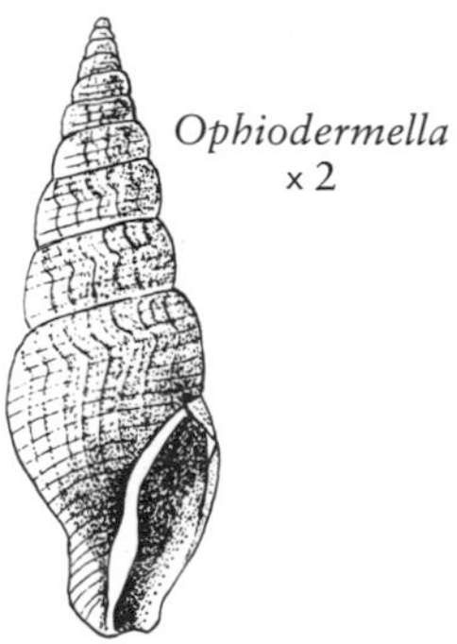

Ophiodermella
× 2

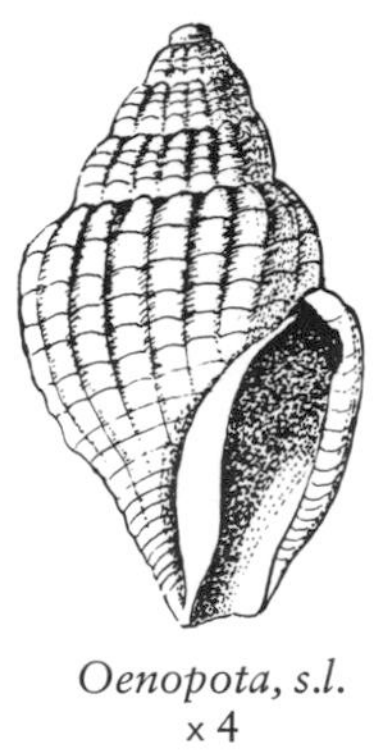

Oenopota, s.l.
× 4

Bellaspira
× 5

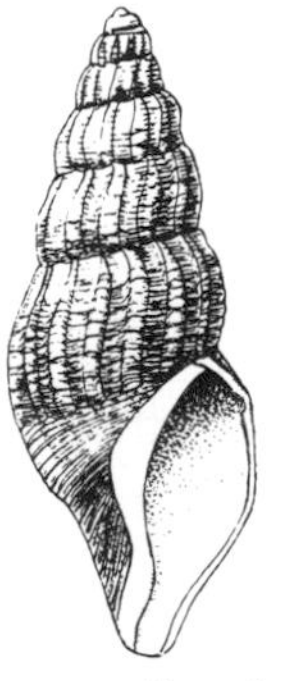

Mangelia, s.l.
× 6

"Pleurotomella"
× 2

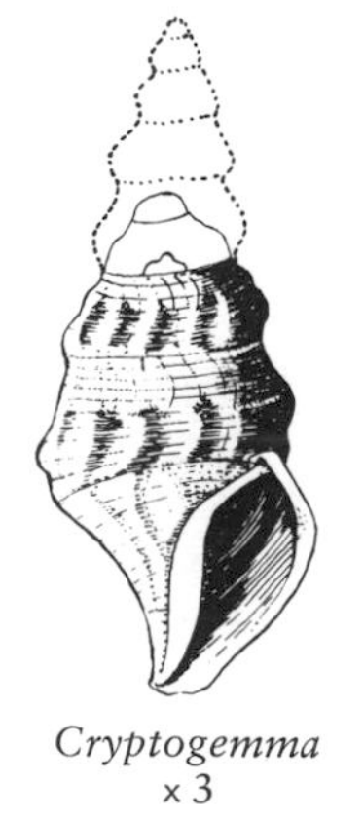

Cryptogemma
× 3

256(255) Sculpture reduced to heavy, well-spaced axial folds .... *Cytharella*
Sculpture of numerous riblets, both axial and spiral ....
.... *Clathurella, s.l.*
257(254) Basal lirae continuing into aperture as low folds .... *Cymakra*
Basal lirae not continuing into aperture 258
258(257) Sculpture of incised spiral grooves and weak axial riblets......259
Sculpture predominantly axial, not smoothly incised 260
259(258) Outer lip contracted anteriorly; axial threads sinuous ....
.... *Ophiodermella*
Outer lip not contracted; axials not sinuous .... *Bellaspira*
260(258) Anterior canal relatively broad .... *Oenopota, s.l.**
Anterior canal relatively narrow 261
261(260) Aperture about half shell length .... *Mangelia, s.l.*
Aperture less than half the shell length 262
262(261) Width of body whorl about half shell height; axial sculpture of oblique riblets .... *"Pleurotomella"*
Width of body whorl less than half shell height; axial sculpture of blunt nodes .... *Cryptogemma*

KEY TO THE PELECYPODA

Orientation of the bivalve shell is, of course, a problem the beginner immediately faces. Key choices make reference to the "right" valve and the "left" valve. The old "rule-of-thumb" guide—that if the closed valves are held in both hands with the beaks pointing in the same direction as one's thumbs, the right valve will be in the right hand and the left valve in the left hand—will work with many clams, but not all. Better guidance to proper orientation is afforded by certain shell structures. Surest guides of all are the soft parts. Both of these sets of criteria are shown in the diagram below, which is the right valve of a generalized clam:

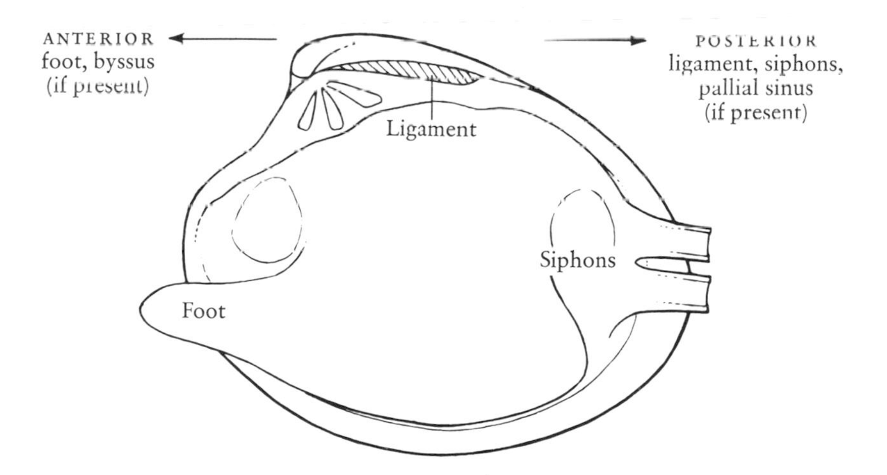

Thus if the closed shell is held in the hands with the ligament area uppermost and nearest to one's body, the right valve will be in the right hand. In shells such as the scallop, *Pecten*, where the ligament is central, the position of the byssal notch gives a clue to orientation, since it is anterior and is more clearly developed in the right valve than in the left. In completely symmetrical shells like *Glycymeris*, where the ligament is distributed in front of as well as behind the beaks, the position of the soft parts (foot and siphon) must be determined for proper orientation. The ligament remains the best external feature for orientation, and to aid in its recognition, most drawings herein show the ligament by diagonal hatching. Finally, we have indicated left and right valves by the letters "L" and "R."

The best feature for the identification of bivalves is the hinge. A complex terminology has been developed for describing the various types of hinge and especially for the exact positions of the hinge teeth in the more highly differentiated forms. Knowledge of this terminology is necessary for the specialist, but, in the interests of simplicity, the terminology is omitted here for the most part. Other and more gross differences are used insofar as possible in the key, leaving the crucial hinge characters for those groups that can be differentiated satisfactorily in no other way. Hinge-tooth patterns have been carefully drawn in the illustrations, and the beginner is advised to study these and the shells themselves and to learn how to make the fine discriminations that are habitual with the experienced worker.

As with gastropods, identification should be made from mature specimens that are reasonably well preserved and unbroken. No key can work infallibly for beachworn or juvenile specimens. Size ranges for adult shells vary so much that the terms denoting size must not be defined or interpreted too rigorously. In general, "minute" shells are less than 5 mm long (i.e. from anterior end to posterior end), "small" shells are 5 to 15 mm, "medium" 15 to 40 mm, and "large" over 40 mm.

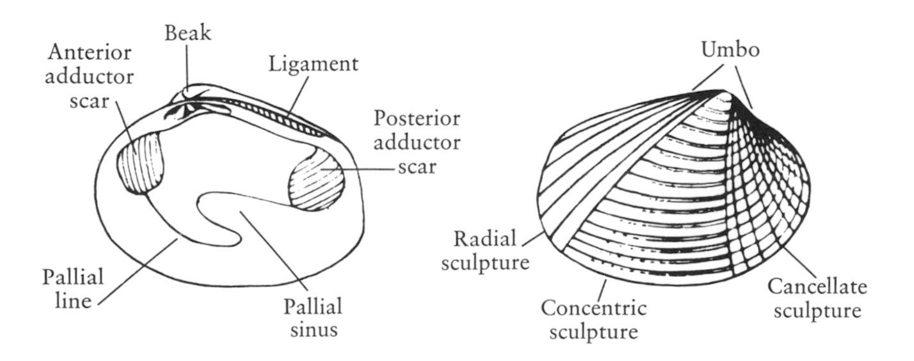

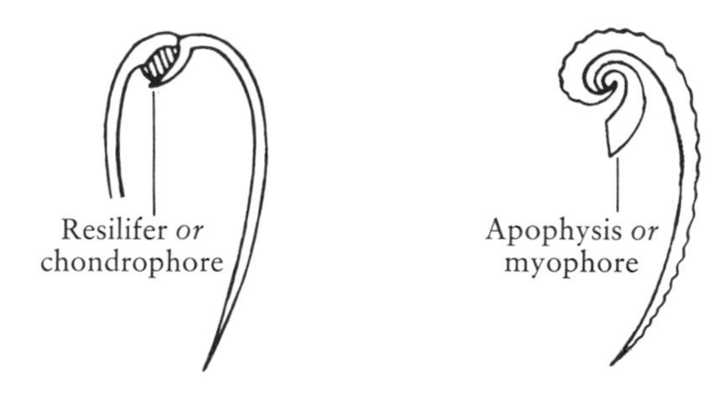

Two types of hinge structure

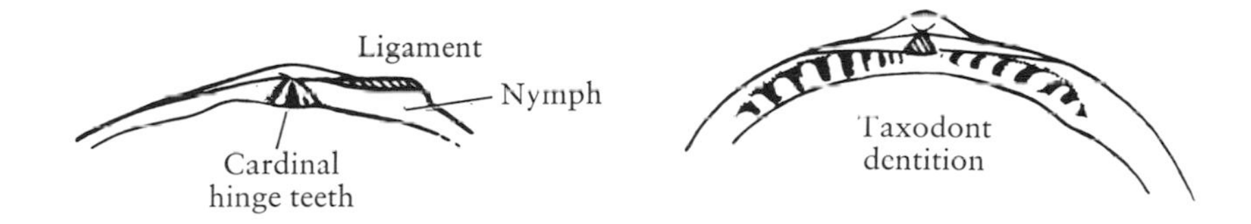

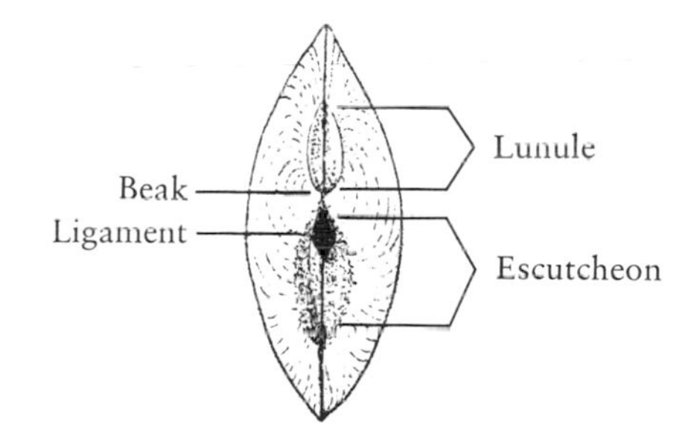

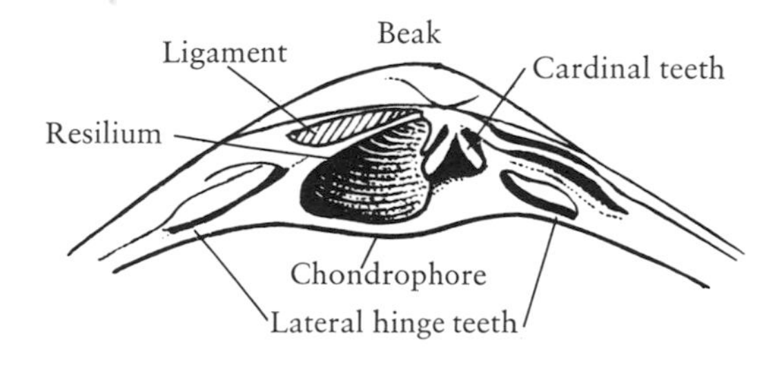

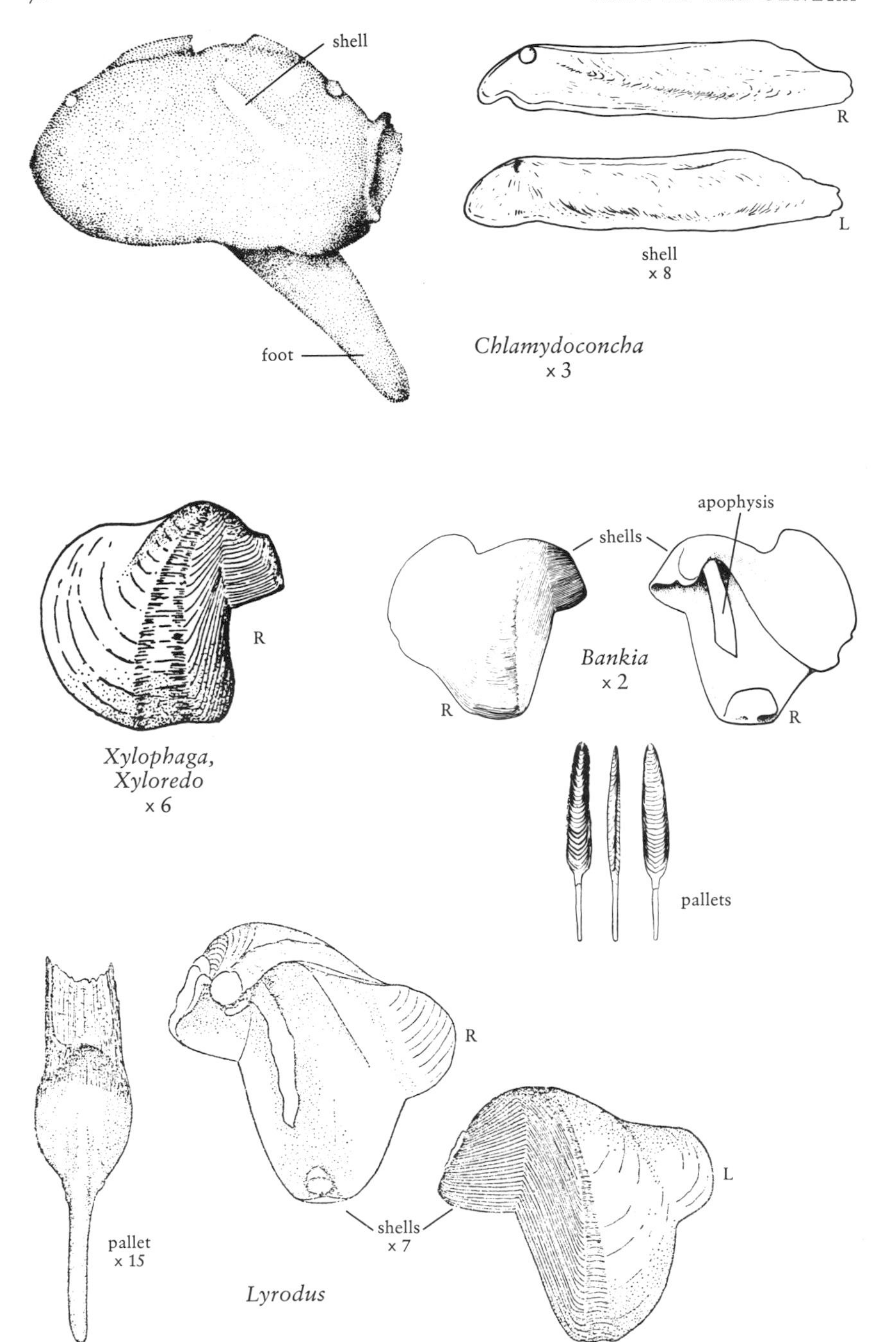
shell
foot
R
L
shell
x 8
Chlamydoconcha
x 3
R
Xylophaga,
Xyloredo
x 6
apophysis
shells
R
Bankia
x 2
R
pallets
R
L
shells
x 7
pallet
x 15
Lyrodus

1 Shell internal, rudimentary; valves not hinged. . . . *Chlamydoconcha*
Shell external, usually covering all or nearly all of soft parts; valves connected by hinge 2

2(1) Valves with an angularly indented or notched anterior margin; animals burrowing in wood 3
Anterior margin evenly arched, mostly convex; animals burrowing in other than wood (mostly rock) 6

3(2) Siphons not equipped with pallets; apophysis lacking . . . . *Xylophaga, Xyloredo*
[*Xyloredo* secreting a calcareous lining in burrow, *Xylophaga* not]
Siphons with protecting pallets; interior of dorsal margin with an apophysis 4

4(3) Pallets of multiple cone-shaped segments . . . . *Bankia*
Pallets paddle-shaped, unitary 5

5(4) Pallets with a brown periostracal cap . . . . *Lyrodus*
Pallets lacking any dark-colored periostracal cap . . . . *Teredo*

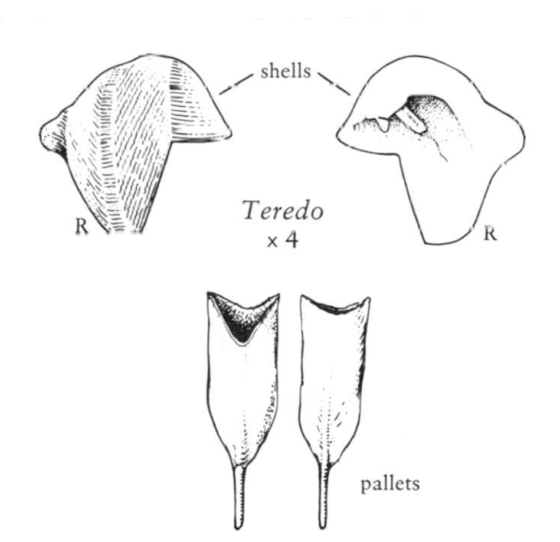

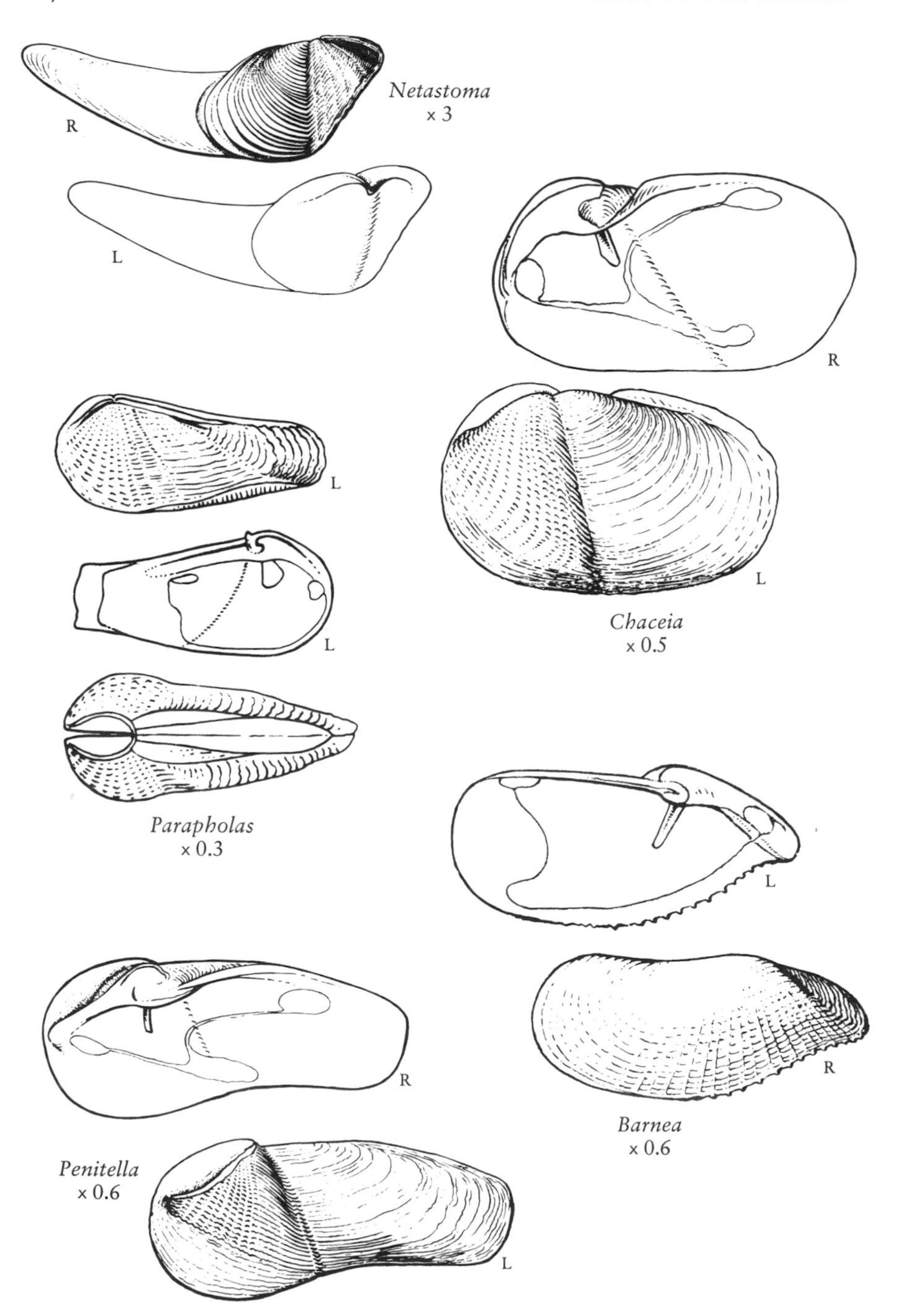

Netastoma × 3

Chaceia × 0.5

Parapholas × 0.3

Barnea × 0.6

Penitella × 0.6

6(2) Anterior end of shell inflated, sculptured by prickly radial ribs; shell with accessory plates 7
Anterior end of shell not inflated, without prickly ribbing; shell without accessory plates 12

7(6) Dorsal margin without internal apophysis; siphonal plates tubular . . . . *Netastoma*
Dorsal margin with internal apophysis; siphonal plates small or lacking, not tubular 8

8(7) Valves divided by sculptural differences into three well-marked regions . . . . *Parapholas*
Valves undivided or with only two areas of different sculpture. . . . 9

9(8) Outline ovoid . . . . *Chaceia*
Shell elongate in outline 10

10(9) Surface not divided into regions . . . . *Barnea*
Surface divided into two clear-cut areas by a sulcus 11

11(10) Posterior gape at end of shell only; apophysis narrow. . . . *Penitella*
Posterior gape extending along dorsal and ventral margins to middle of shell; apophysis broad . . . . *Zirfaea*

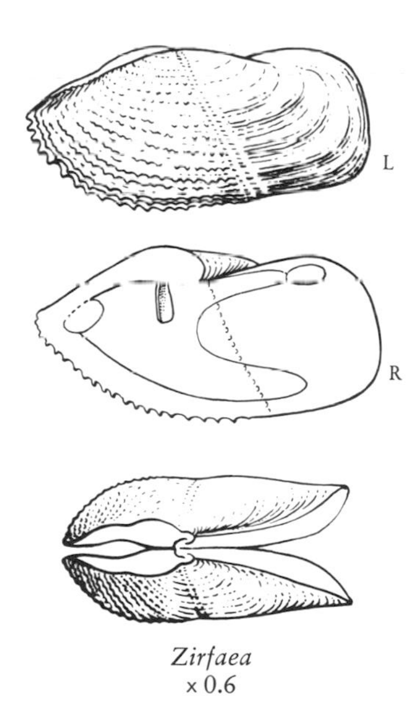

Zirfaea
x 0.6

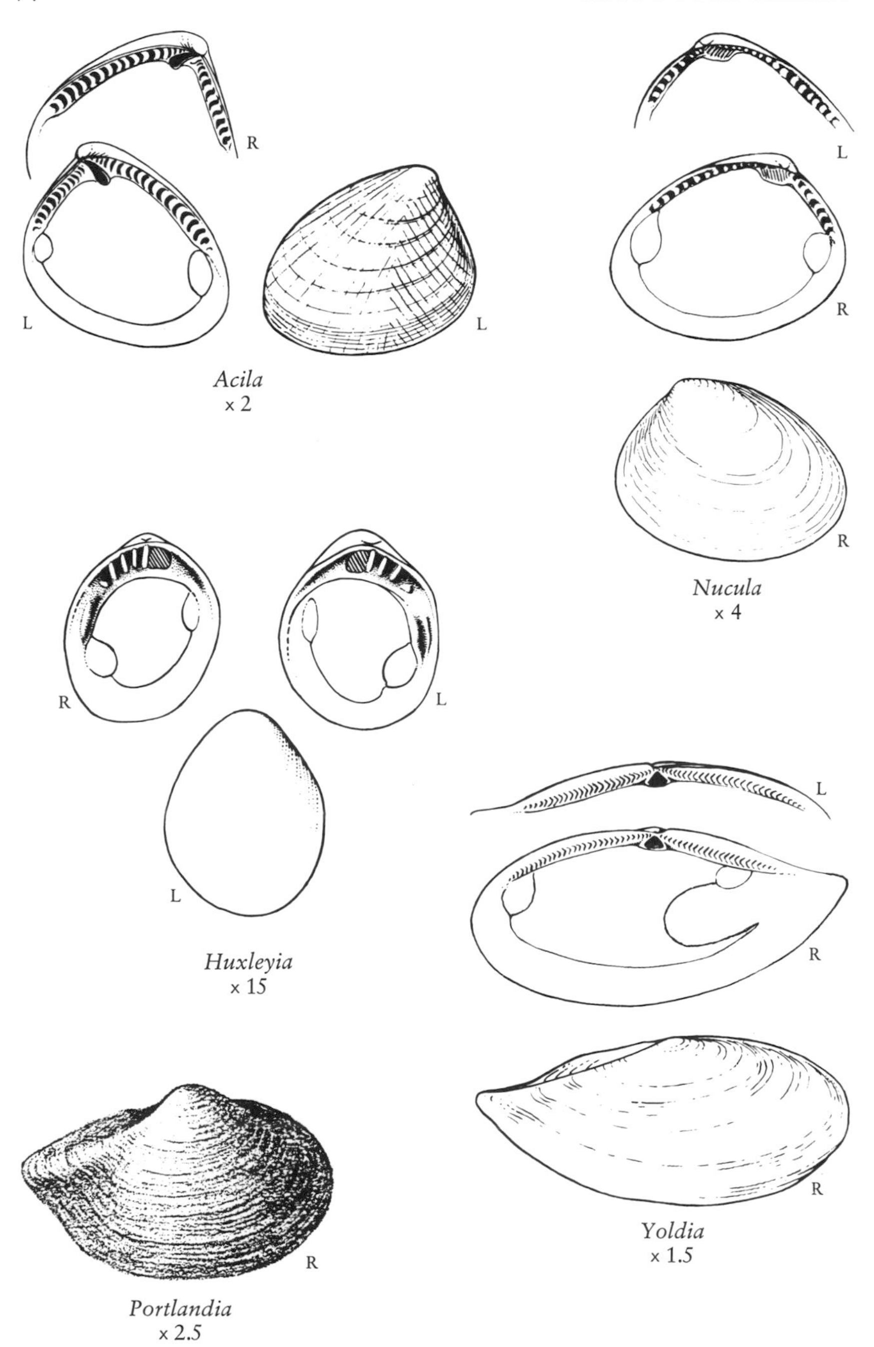

Acila
x 2

Nucula
x 4

Huxleyia
x 15

Yoldia
x 1.5

Portlandia
x 2.5

12(6) Hinge with taxodont dentition 13
Hinge with other than taxodont dentition 28
13(12) Interior of valves nacreous 14
Interior of valves porcelaneous or at most subnacreous 15
14(13) Sculpture of zigzag riblets . . . . *Acila*
Sculpture wanting or of even-sized radial riblets . . . . *Nucula*
15(13) Sculpture obsolete or predominantly concentric (rarely, with faint radial striae) 16
Sculpture predominantly radial (rarely, with faint concentric striae) 23
16(15) Hinge plate markedly broad; hinge teeth few . . . . *Huxleyia*
Hinge plate not markedly broad; teeth numerous 17
17(16) Valves slightly gaping; ligament in a broad internal pit; pallial sinus wide and deep . . . . *Yoldia*
Valves tightly closing; ligamental pit narrow or ligament external; pallial sinus narrow, shallow, or indistinct 18
18(17) Posterior margin rostrate or pointed 19
Posterior margin evenly rounded 20
19(18) Posterior rostration set off by a furrow; shell surface smooth or nearly so . . . . *Portlandia*
Posterior rostration not set off by a furrow; shell surface with concentric sculpture . . . . *Nuculana*

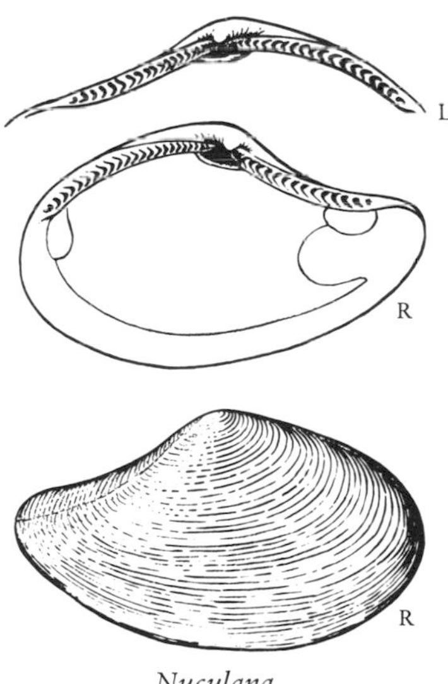

Nuculana
x 3

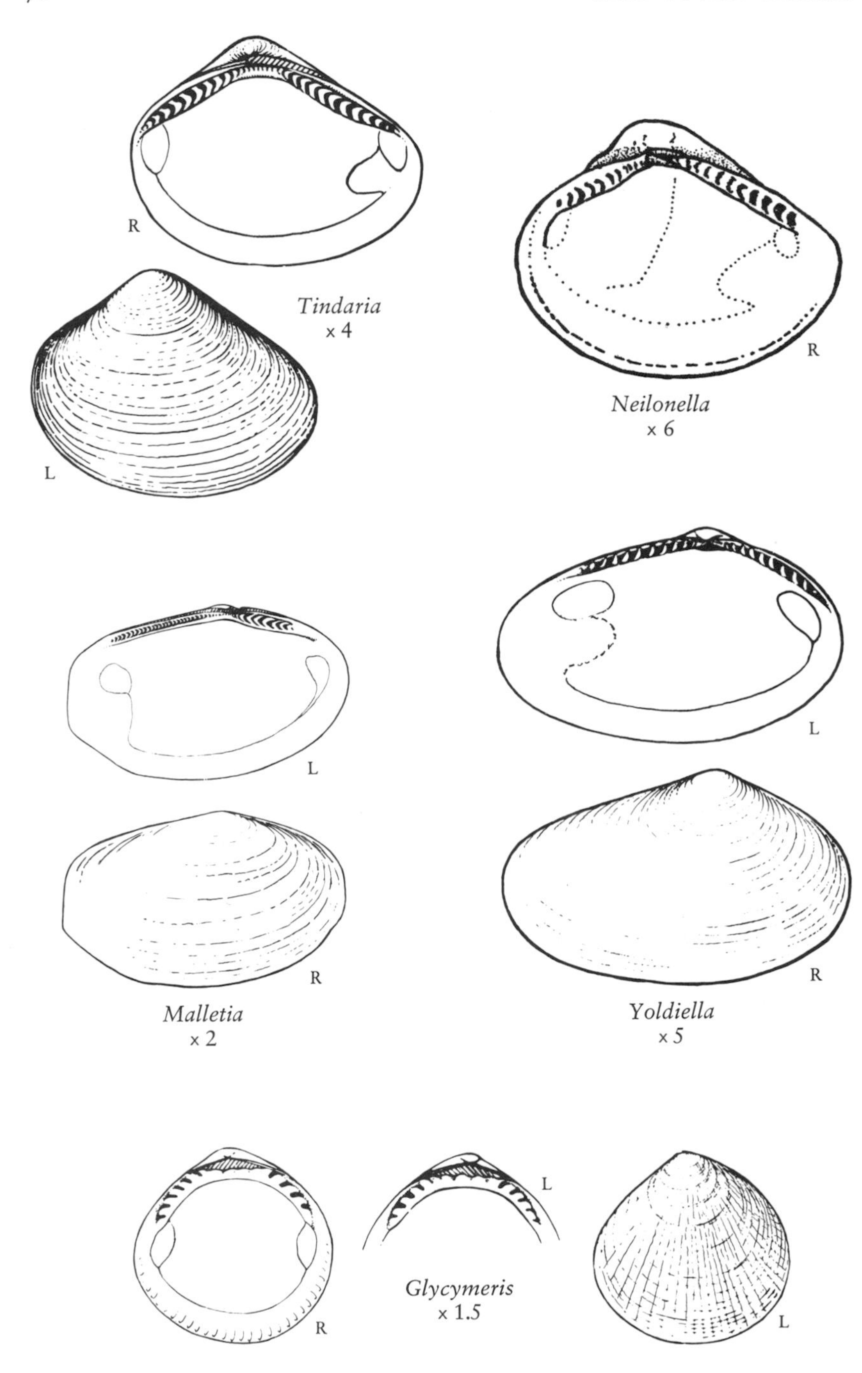
R
L
Tindaria
x 4
R
Neilonella
x 6
L
R
Malletia
x 2
L
R
Yoldiella
x 5
R
L
Glycymeris
x 1.5
L

20(18) Ligament partially internal . . . . *Neilonella*
Ligament entirely external 21

21(20) Shell ovate, thick; beaks prominent . . . . *Tindaria*
Shell somewhat quadrate, thin; beaks not prominent 22

22(21) Teeth of posterior hinge section markedly more numerous than in anterior part of hinge; pallial sinus distinct . . . . *Malletia*
Teeth of two sections of hinge not markedly different in number; pallial sinus, if present, indistinct . . . . *Yoldiella*

23(15) Hinge line arched . . . . *Glycymeris*
Hinge line relatively straight, not arched 24

24(23) Ligament in a deep central pit visible from above . . . . *Limopsis*
Ligament not confined to a central pit 25

25(24) Sculpture of coarse ribs, grooved or noded; ligamental area wide, with chevron-shaped grooves . . . . *Anadara*
Sculpture of fine radial ribs and faint concentric lines; ligamental area narrow 26

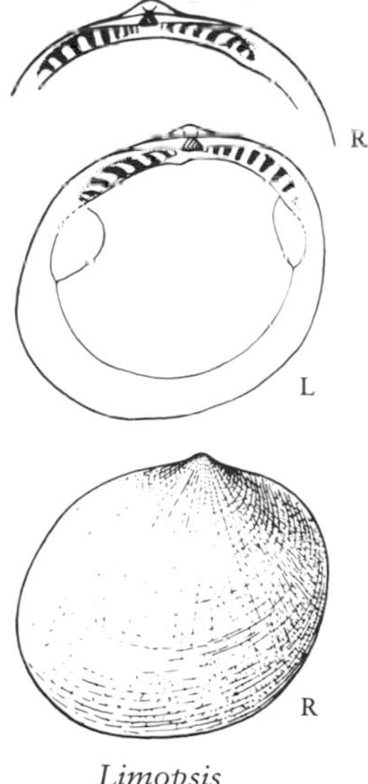

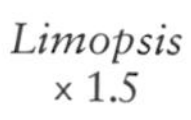

Limopsis
× 1.5

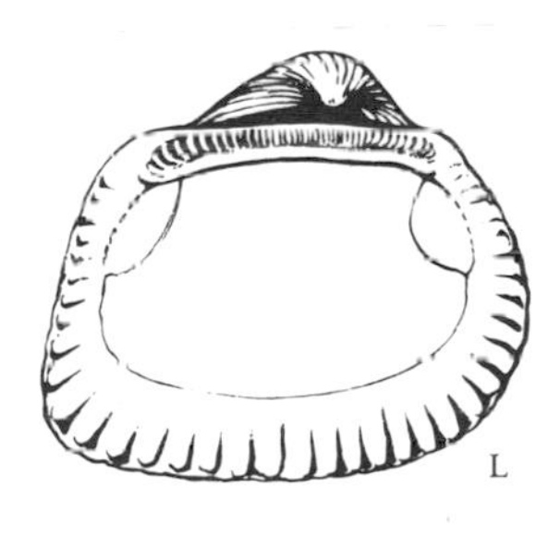

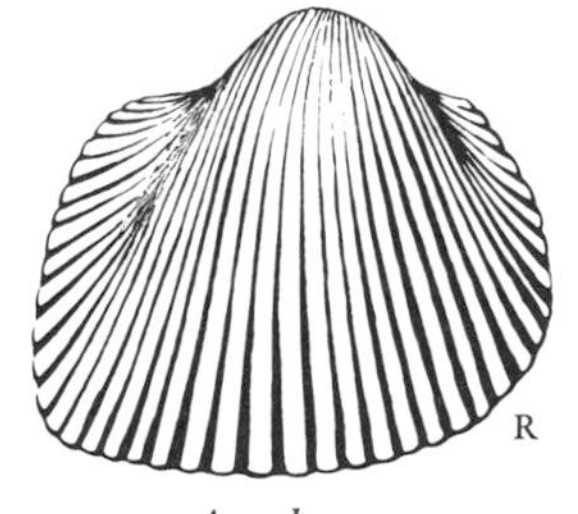

Anadara
× 0.6

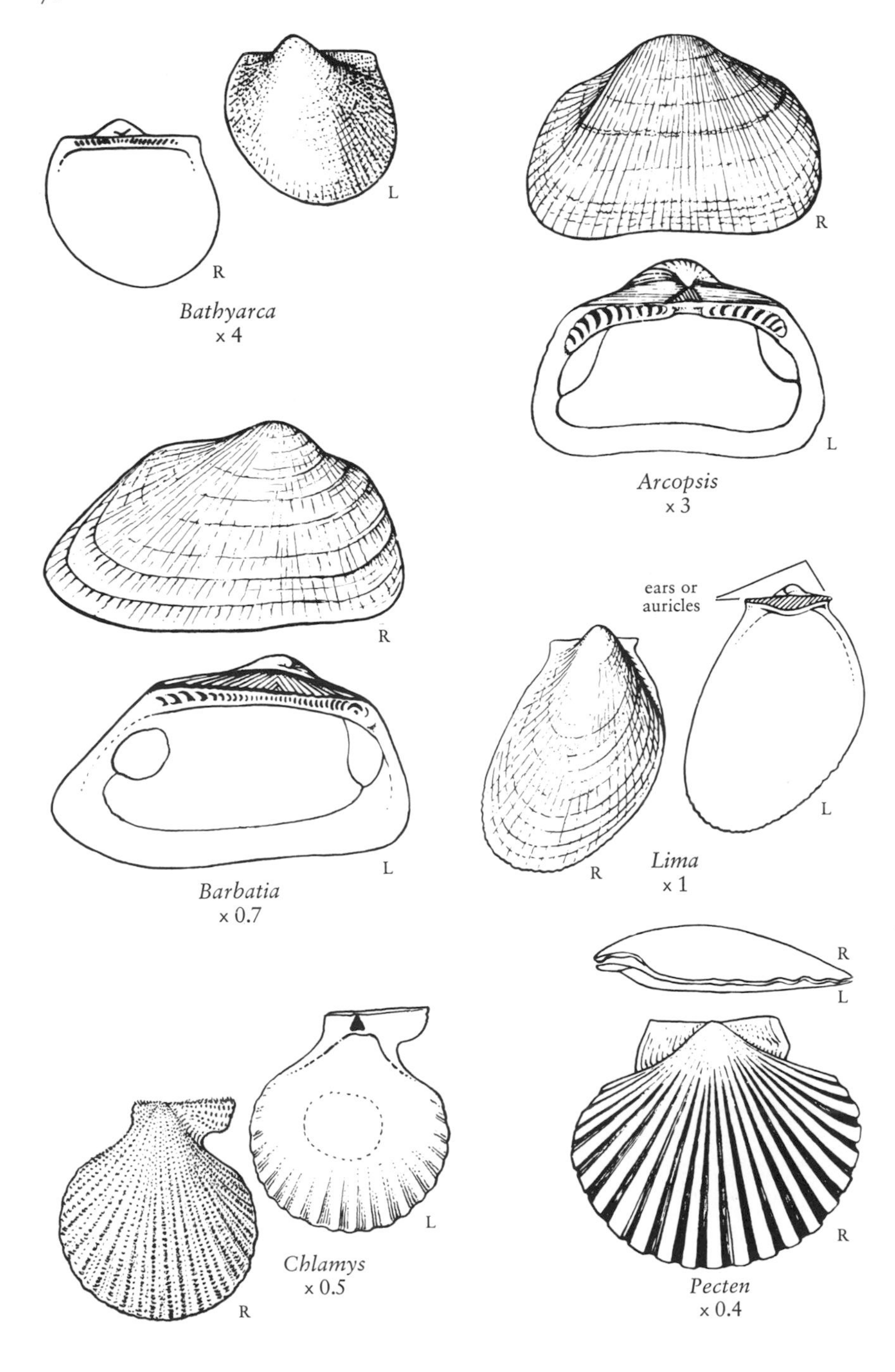

Bathyarca
x 4

Arcopsis
x 3

Barbatia
x 0.7

Lima
x 1

Chlamys
x 0.5

Pecten
x 0.4

26(25) Hinge teeth relatively few; shell thin and light .... *Bathyarca*
Hinge teeth numerous; shell sturdy 27
27(26) Ligamental area extending along full length of dorsal margin .... *Barbatia*
Ligamental area contracted, diamond-shaped .... *Arcopsis*
28(12) Dorsal margin produced anteriorly and posteriorly into triangular ears (auricles) 29
Dorsal margin not produced or eared 38
29(28) Shell distinctly higher than long; ears small, not well set off... *Lima*
Shell height nearly the same as the length or less than length; at least one ear well set off by a furrow 30
30(29) With radial ribs relatively well-developed, corrugating the shell 31
With weak or delicate radial ribbing, not corrugating the shell... 36
31(30) Growth stages evident, juvenile shell regular, adult distorted by attachment; hinge with a purple stain .... *Hinnites*
Growth stages not evident, surface undistorted; hinge without purple stain 32
32(31) Anterior ears more than twice as long as posterior ...... *Chlamys*
Anterior ears less than twice as long as posterior or approximately equal in size 33
33(32) Valves of unequal convexity, right valve noticeably more convex, left flat to concave .... *Pecten*
Valves of nearly equal convexity 34
34(33) Shell thin; valves only slightly convex .... *Leptopecten*
Shell sturdy in texture; valves strongly convex 35

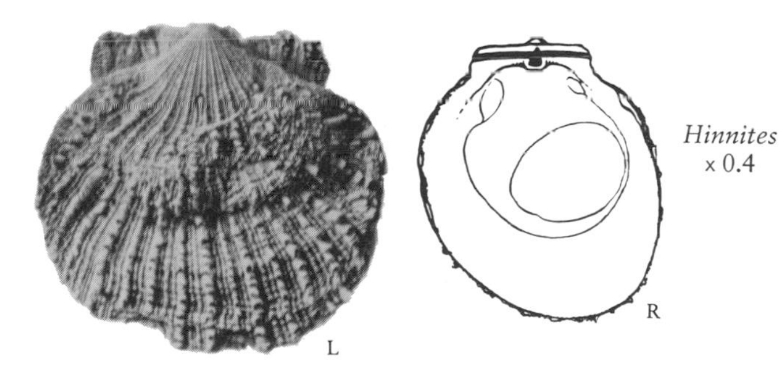

Hinnites
x 0.4

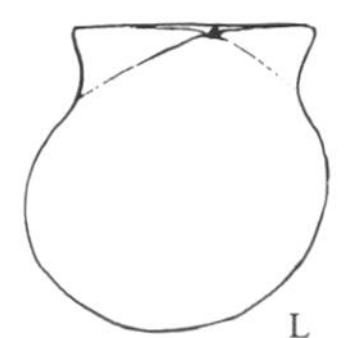

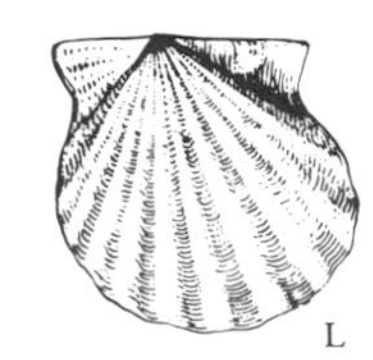

Leptopecten
x 1

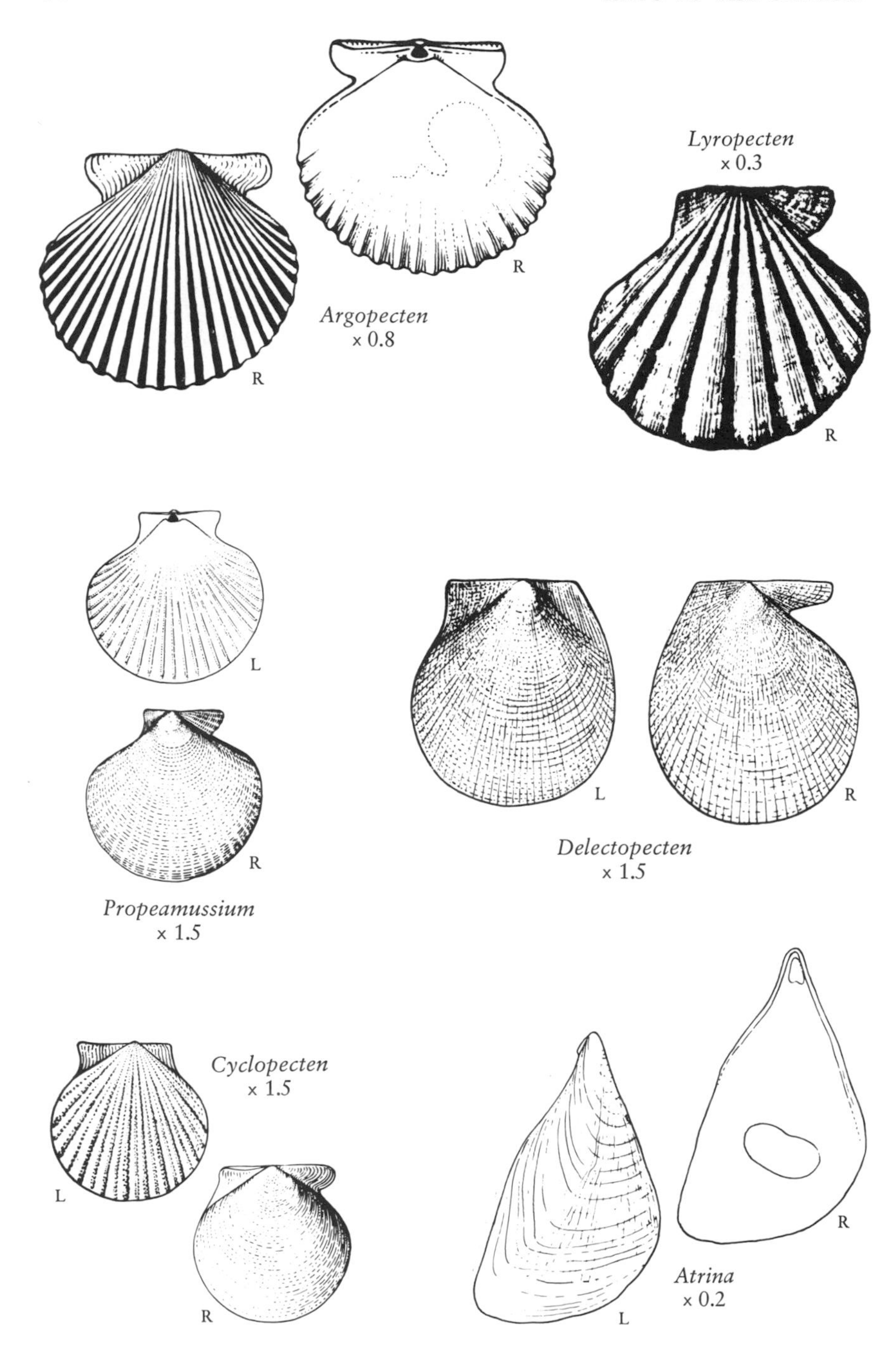

Argopecten × 0.8

Lyropecten × 0.3

Propeamussium × 1.5

Delectopecten × 1.5

Cyclopecten × 1.5

Atrina × 0.2

35(34) Ribs simple, without subdivision . . . . *Argopecten*
Ribs subdivided by small, overlying riblets . . . . *Lyropecten*
36(30) Exterior smooth or with fine threads, interior with slender ribs
. . . . *Propeamussium*
Exterior markedly ribbed on one or both valves 37
37(36) Right valve with concentric, left with radial ribs . . . . . *Cyclopecten*
Both valves with radial or reticulate sculpture . . . . . . *Delectopecten*
38(28) Shell large, triangular, dark-colored, with a central nacreous area within . . . . *Atrina*
Shell not large, triangular and dark-colored 39
39(38) Adductor scars coalesced, appearing as one large scar near center of shell 40
Adductor scars separated, at opposite ends of shell 44
40(39) Adductor scar complex, central area showing superimposed secondary scars 41
Adductor scar simple, without secondary scars 42
41(40) Central muscle scar area having one or two smaller scars within
. . . . *Pododesmus*
Central muscle scar having three smaller scars . . . . *Anomia*
42(40) Shell minute, ligament elongate . . . . *Philobrya*
Shell medium-sized to large; ligament trigonal 43

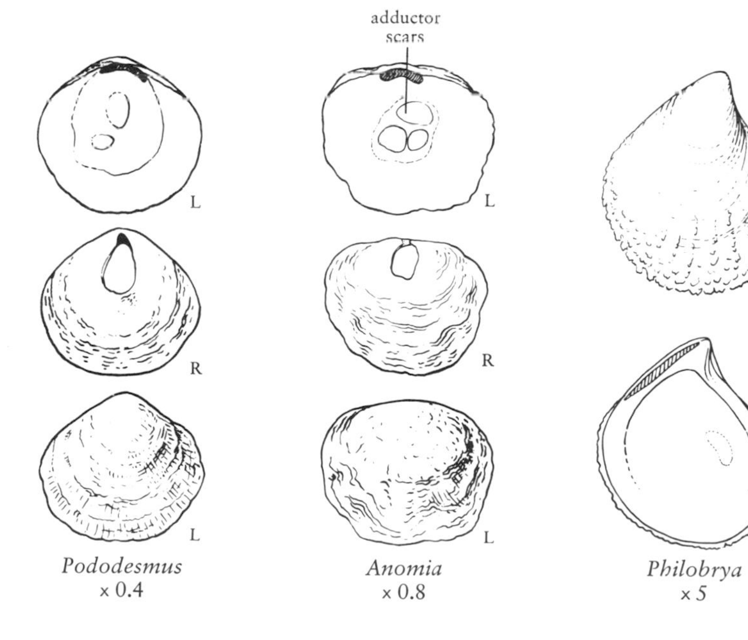

Pododesmus × 0.4 | *Anomia* × 0.8 | *Philobrya* × 5

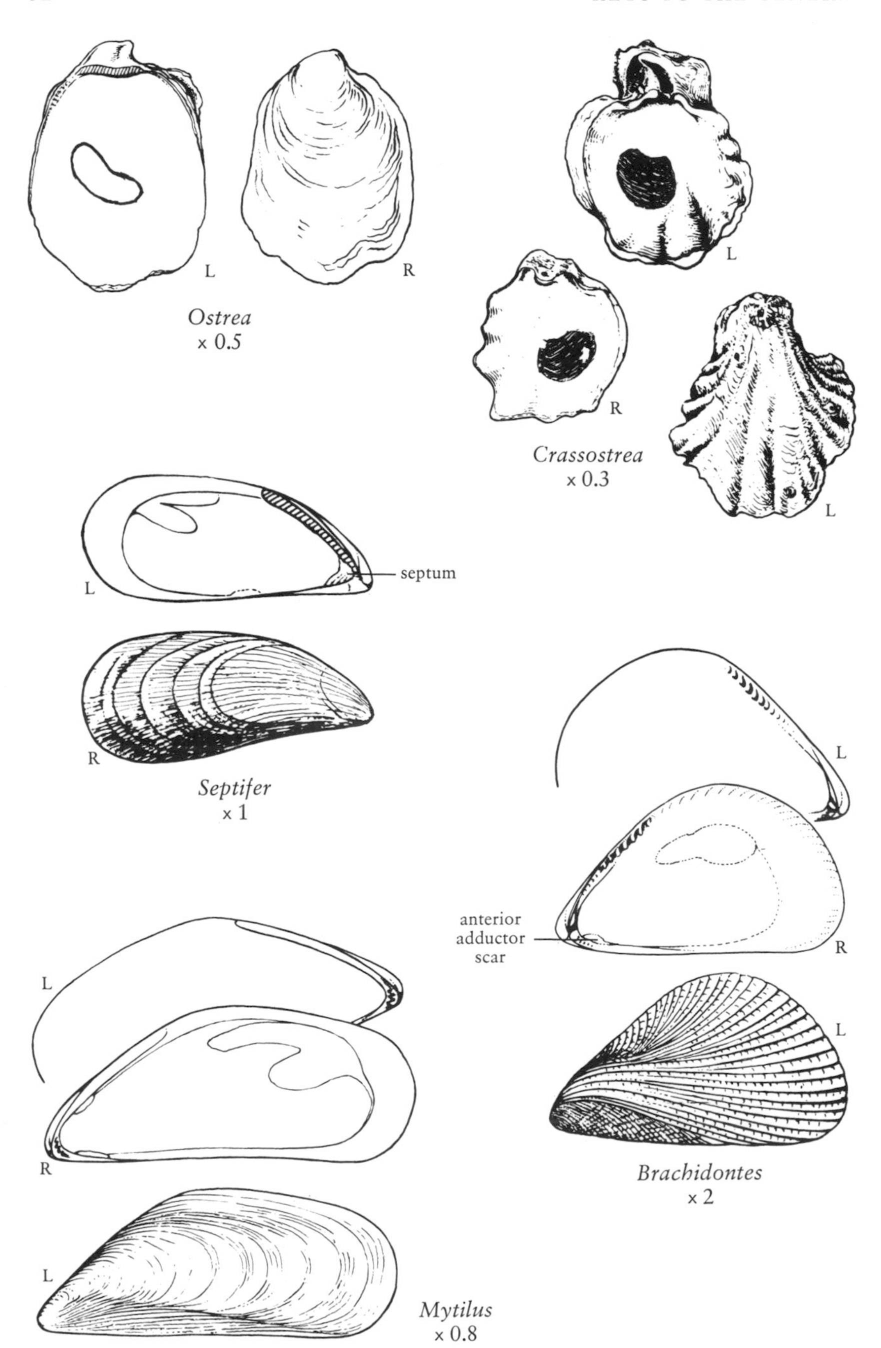

Ostrea
x 0.5

Crassostrea
x 0.3

Septifer
x 1

Brachidontes
x 2

Mytilus
x 0.8

43(42) Inner margin with fine crenulations near hinge; muscle scar nearly central, kidney-shaped . . . . *Ostrea*
Inner margin smooth; muscle scar near posterior margin, elongate, squared above, often dark-colored . . . . . . *Crassostrea*

44(39) Hinge plate without true teeth or other interlocking or conspicuous lamellar projections (irregular denticles and nonprojecting resilifers may be present) 45
Hinge plate with projecting or interlocking teeth or with a projecting chondrophore 74

45(44) Adductor muscle scars very unequal in size; anterior scar smaller, near beaks 46
Adductor muscle scars approximately equal in size (not necessarily in shape), nearly equidistant from beaks 58

46(45) Beaks terminal, at anterior end of shell 47
Beaks near anterior end but not terminal 49

47(46) Anterior end bridged within by a shelly septum . . . . *Septifer*
Anterior end open within 48

48(47) Shell with regular radial ribs, markedly finer anteriorly and ventrally; hinge smooth . . . . *Brachidontes*
Shell smooth or with a few coarse radial ribs (ribs not finer ventrally); hinge with some small denticles near beaks . . . *Mytilus*

49(46) Shell cylindrical (i.e. elongate, with dorsal and ventral margins parallel) 50
Shell not cylindrical 52

50(49) Periostracum studded with hairs; lunule with radial sculpture . . . . *Gregariella*
Periostracum roughened but not hairy; lunule smooth 51

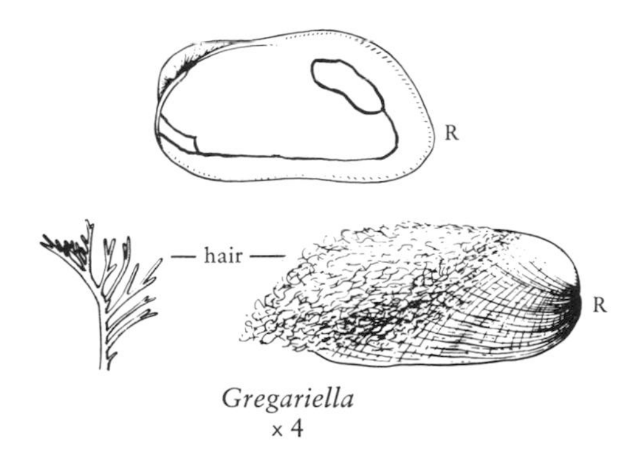

Gregariella
x 4

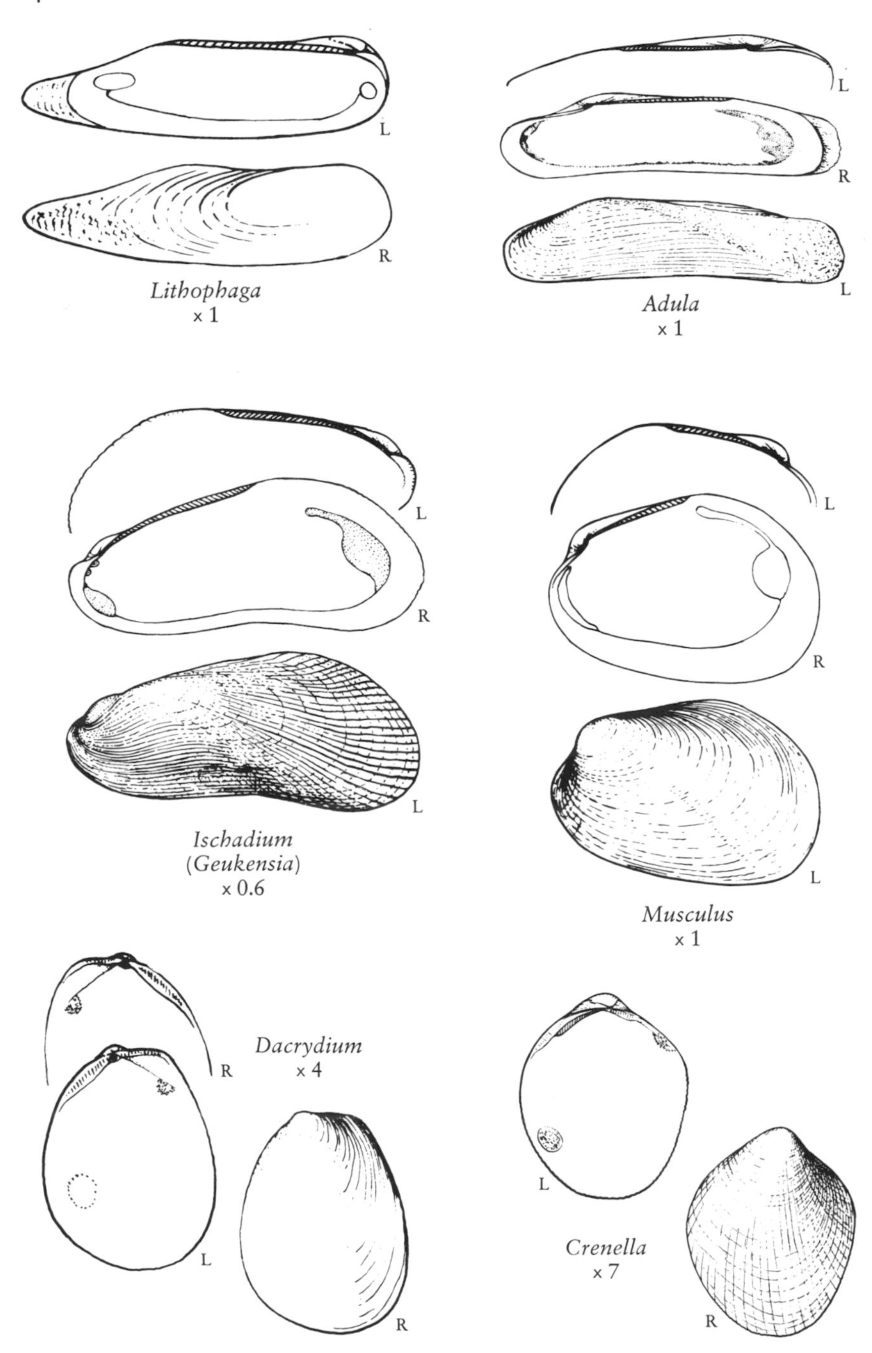

Lithophaga
× 1

Adula
× 1

Ischadium
(*Geukensia*)
× 0.6

Musculus
× 1

Dacrydium
× 4

Crenella
× 7

51(50) Posterior-dorsal slope coated with rough, chalky incrustations; inner margins smooth .... *Lithophaga*
Posterior-dorsal slope without rough, chalky incrustations; dorsal margin crenulate in some .... *Adula*

52(49) With radial ribbing at one or both ends of shell, middle smoother 53
Shell entirely smooth or with relatively even ribbing over entire surface 54

53(52) Radial ribs coarser posteriorly .... *Ischadium (Geukensia)*
Radial ribs coarser anteriorly .... *Musculus*

54(52) Shell ovate, small to minute 55
Shell elongate, posterior end flaring, medium-sized to large 57

55(54) Shell smooth, nearly transparent; hinge with some denticles .... *Dacrydium*
Shell with divaricate radial ribbing; hinge smooth or with a resilifer 56

56(55) Radial ribs relatively strong; ligament sunken, in a toothlike, crenulated resilifer .... *Crenella*
Radial ribs weak; ligament nearly marginal, not in a toothlike resilifer .... *Megacrenella*

57(54) Periostracum smooth, shiny, never hairy .... *Amygdalum*
Periostracum opaque, brownish, posterior slope sparsely to densely hairy .... *Modiolus*

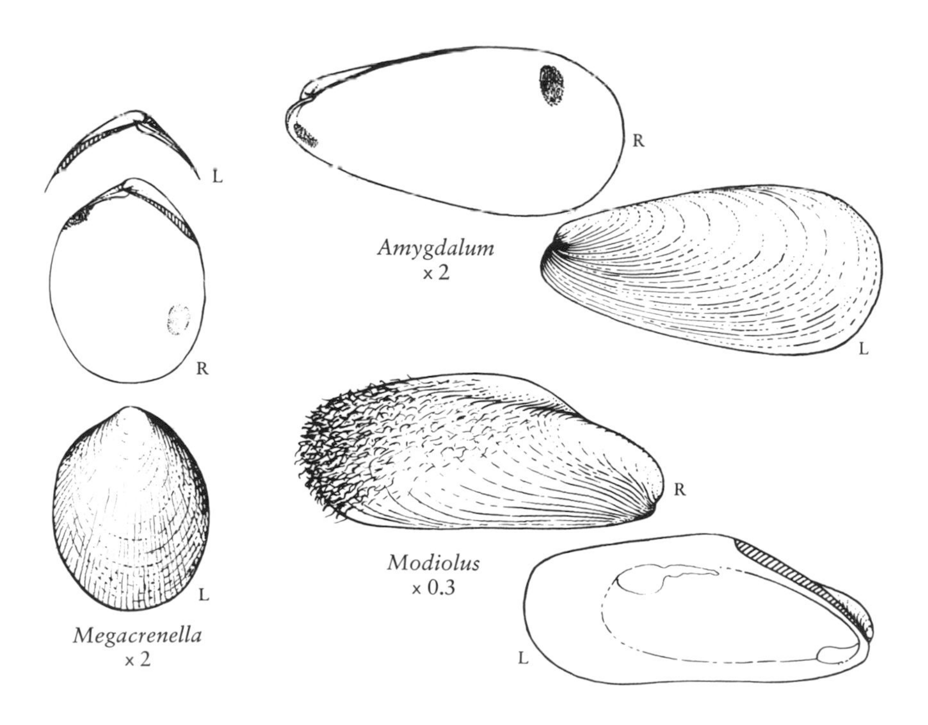

Megacrenella x 2

Amygdalum x 2

Modiolus x 0.3

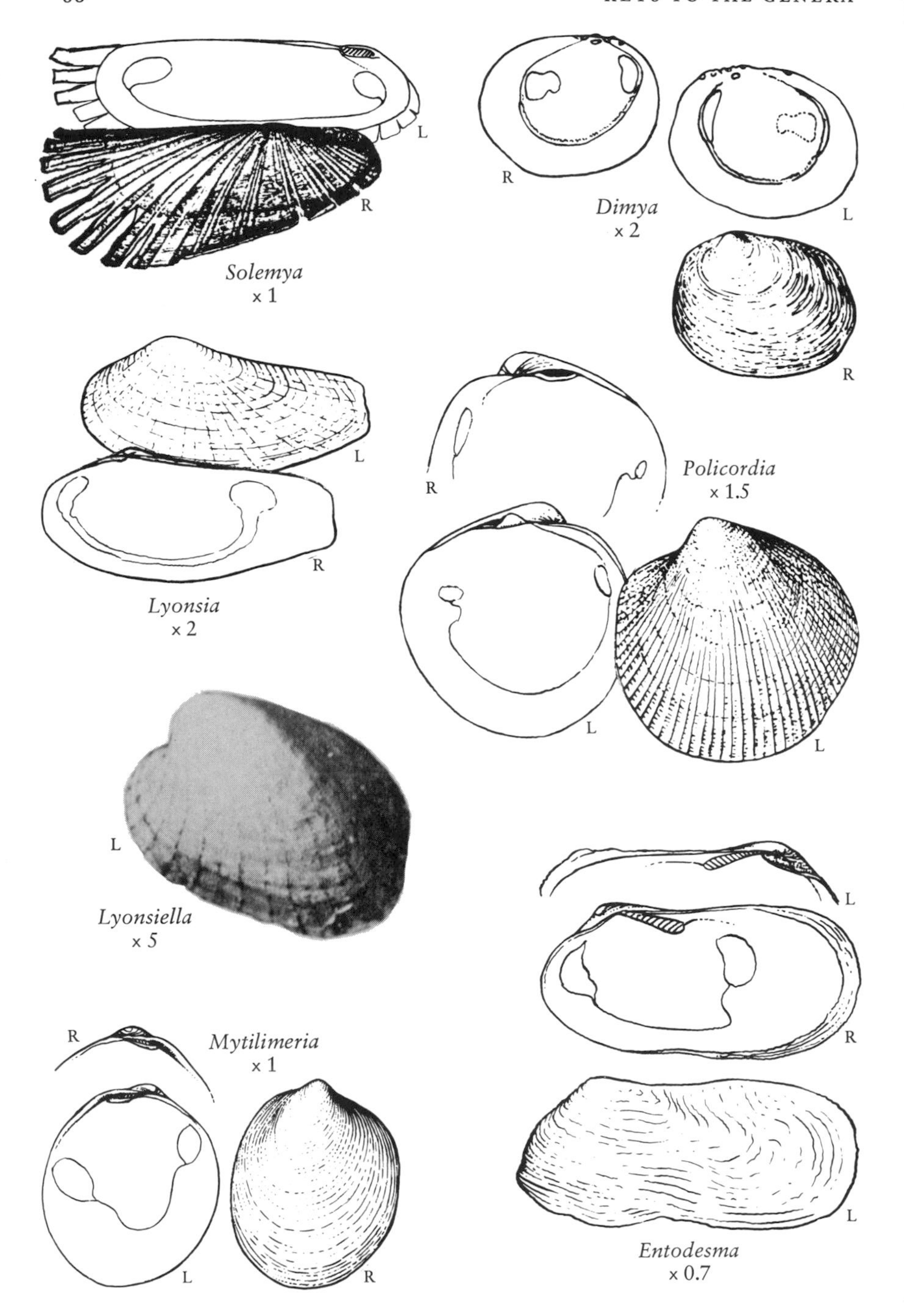

Solemya × 1

Dimya × 2

Lyonsia × 2

Policordia × 1.5

Lyonsiella × 5

Mytilimeria × 1

Entodesma × 0.7

58(45) Periostracum prolonged as a fringe beyond margin of shell .... *Solemya*
Periostracum, if present, not prolonged beyond margin 59
59(58) Shell material nacreous or pearly, especially within 60
Shell material porcelaneous or chalky, not nacreous 64
60(59) Outline subovate to nearly circular 61
Posterior end more or less drawn out 64
61(60) Compressed; sculpture of concentric laminae; right valve attached to hard substrate .... *Dimya*
Inflated; surface smooth or radially ribbed 62
62(61) Surface smooth .... *Mytilimeria*
Surface with radial sculpture 63
63(62) Shell subquadrate; radial riblets well spaced .... *Lyonsiella*
Shell subcircular; radial riblets fine .... *Policordia*
64(60) Periostracum thin, sculptured in fine radial lines .... *Lyonsia*
Periostracum heavy, radial lines coarse or irregular .... *Entodesma*
65(59) Shell gaping, either ventrally or anteriorly 66
Shell not gaping, valves closing tightly 67
66(65) Periostracum heavy, dark-colored; shell rectangular, beaks central .... *Cyrtodaria*
Periostracum thin and light-colored, if present; shell oblique-ovate to asymmetrical .... *Hiatella**

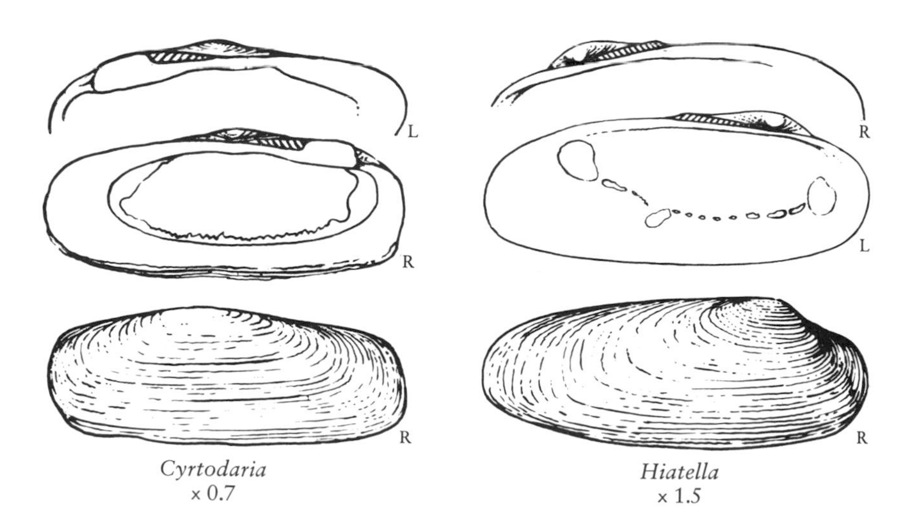

Cyrtodaria
× 0.7

Hiatella
× 1.5

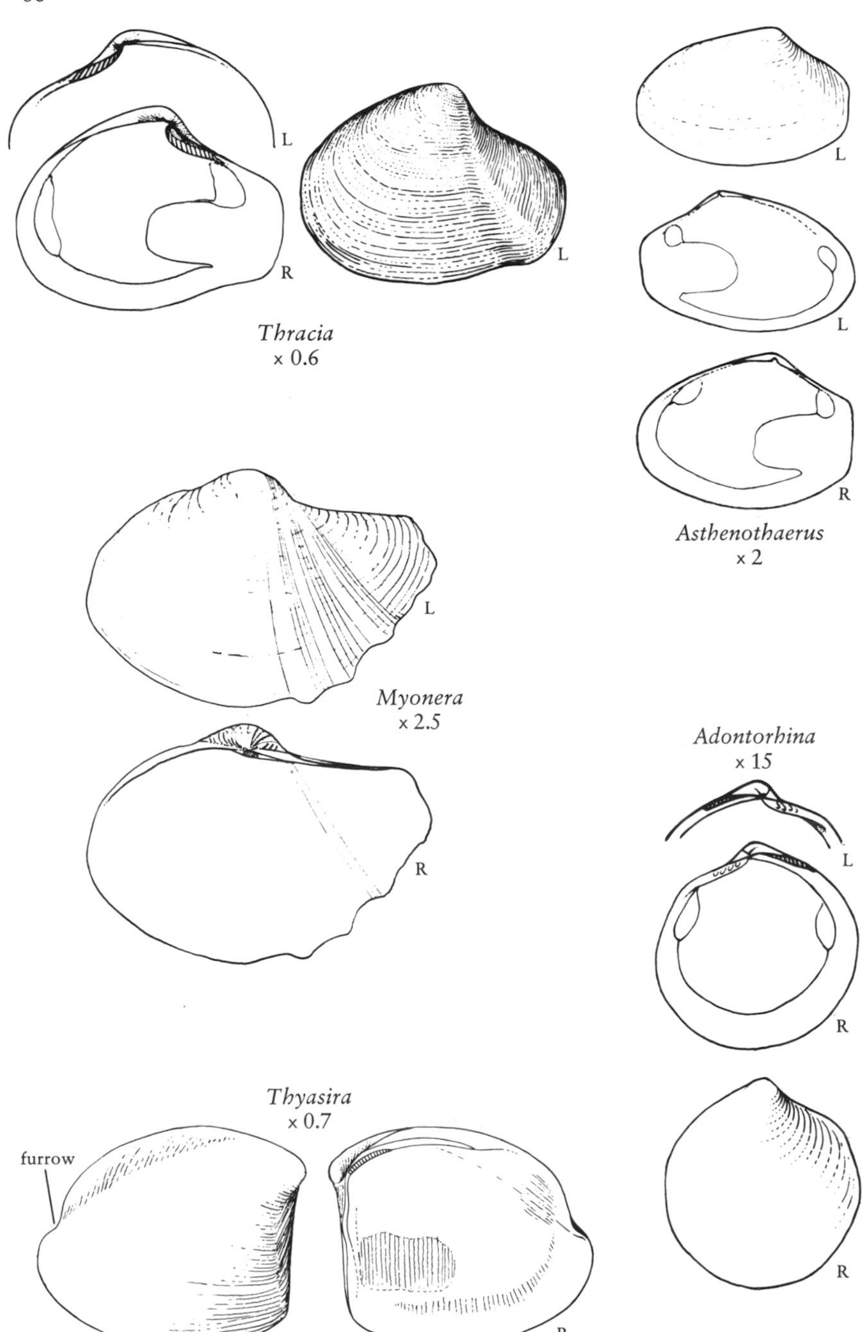

Thracia
× 0.6

Asthenothaerus
× 2

Myonera
× 2.5

Adontorhina
× 15

Thyasira
× 0.7

67(65) With conspicuous pallial sinus 68
With no sinus, pallial line entire or only wavy 69

68(67) Ligament external .... *Thracia*
Ligament internal, concealed below umbones .... *Asthenothaerus*

69(67) With one or more conspicuous furrows setting off posterior slope 70
Posterior slope or area not bounded by a furrow 71

70(69) Dorsal margin long, posterior end spoutlike .... *Myonera*
Dorsal margin convex, posterior end not spoutlike ..... *Thyasira*

71(69) Subcircular in outline, inflated; length less than 5 mm 72
Trigonal to quadrate or trapezoidal in outline, not inflated; length more than 5 mm 73

72(71) Hinge plate with several irregular denticles .... *Adontorhina*
Hinge plate smooth, laminar .... *Tomburchus**

73(71) Shell trapezoidal in outline, smooth, hinge with a central pit in one valve .... *Saxicavella*
Shell trigonal, with radial ribs at ends, no pit on hinge plate .... *Serripes**

74(44) Shell gaping at both ends, cylindrical, height about two-fifths length 75
Shell not gaping at both ends, not cylindrical 78

75(74) Beaks at or near anterior end 76
Beaks subcentral 77

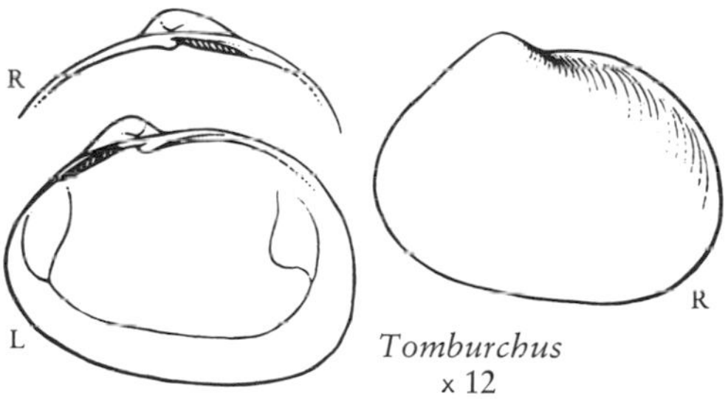

Tomburchus
x 12

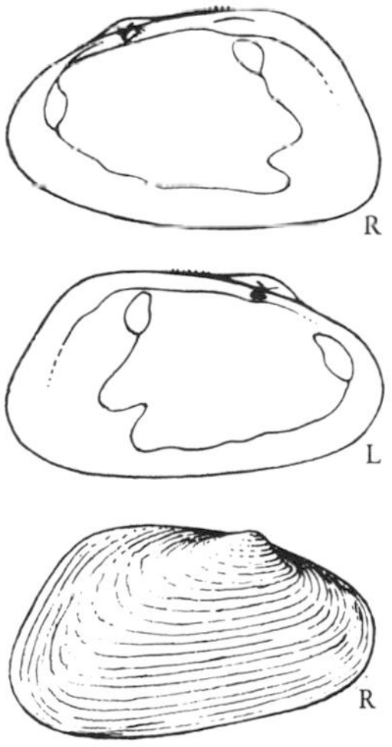

Saxicavella
x 5

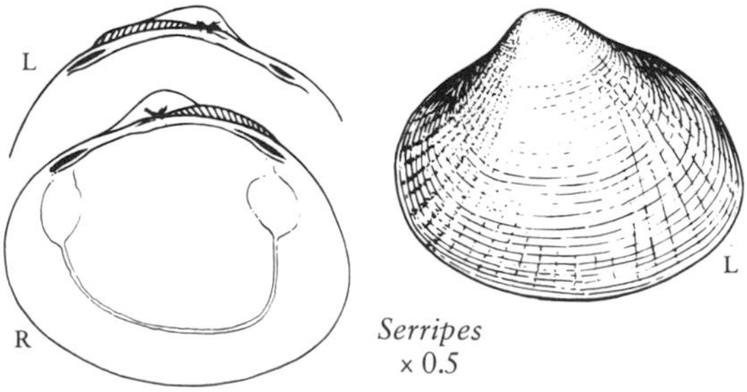

Serripes
x 0.5

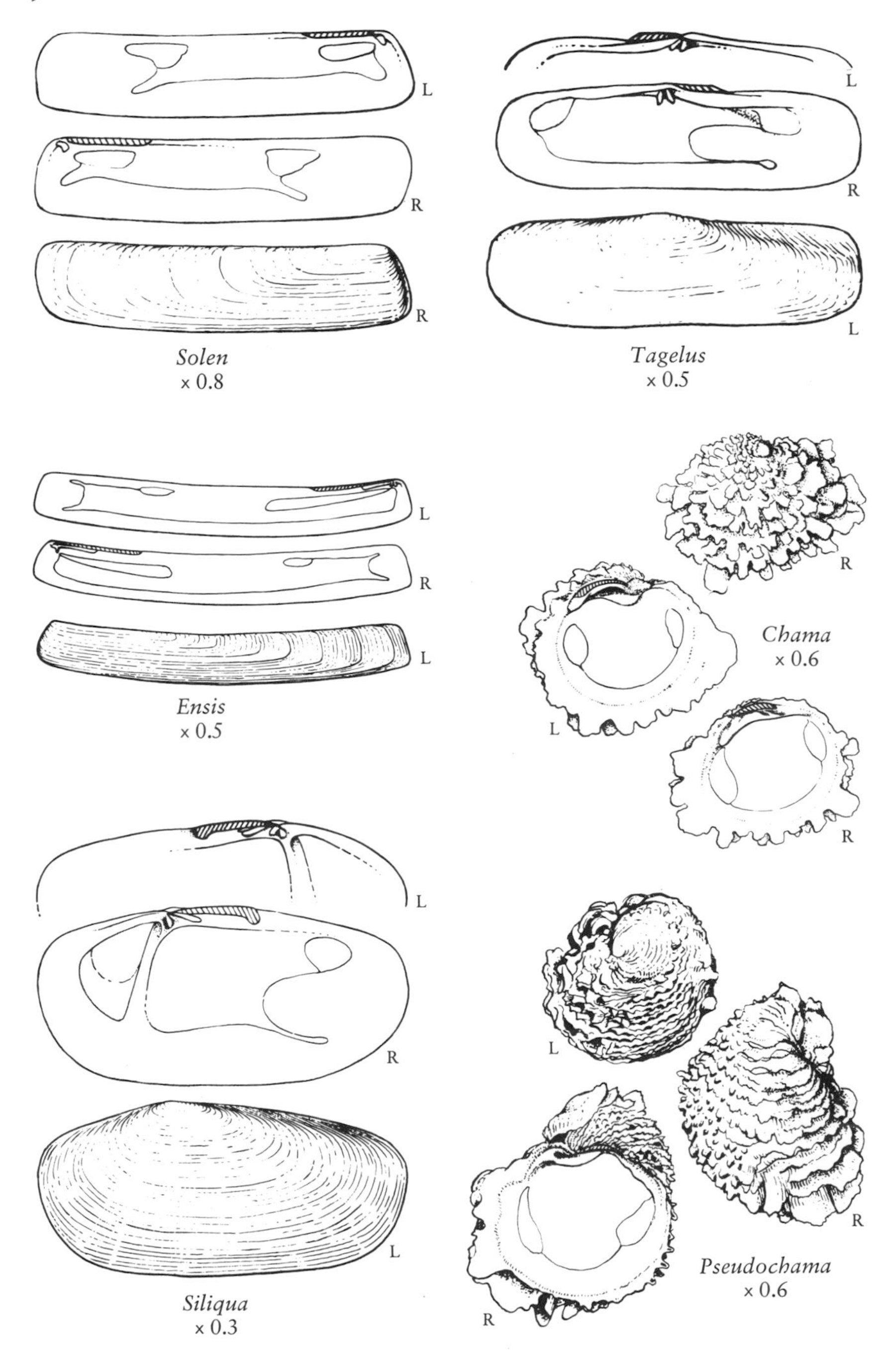

Solen
× 0.8

Tagelus
× 0.5

Ensis
× 0.5

Chama
× 0.6

Siliqua
× 0.3

Pseudochama
× 0.6

76(75) Dorsal margin nearly straight, not arched . . . . *Solen*
Dorsal margin curved or slightly arched . . . . *Ensis*
77(75) Interior with a vertical rib . . . . *Siliqua*
Interior of shell without a reinforcing rib . . . . *Tagelus*
78(74) Shell attached by one valve, beaks spirally twisted, sculpture of irregular concentric lamellae 79
Shell not attached, beaks not twisted, sculpture various but not of irregular lamellae 80
79(78) Attachment by left valve . . . . *Chama*
Attachment by right valve . . . . *Pseudochama*
80(78) Posterior end drawn out into a spout or beak 81
Posterior end evenly rounded 83
81(80) Sculpture of radial ribs . . . . *Cardiomya*
Sculpture not radial 82
82(81) Surface smooth . . . . *Cuspidaria*
Sculpture of granules over entire surface . . . . *Plectodon*

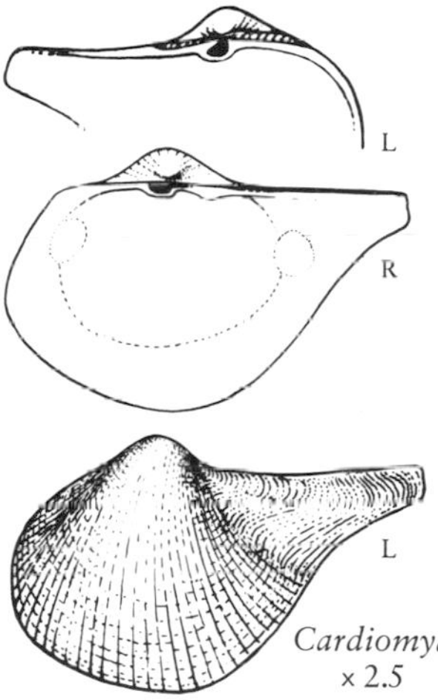

Cardiomya
× 2.5

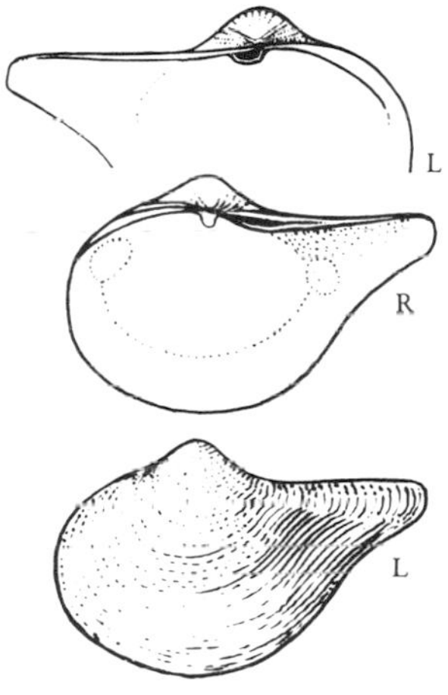

Cuspidaria
× 1

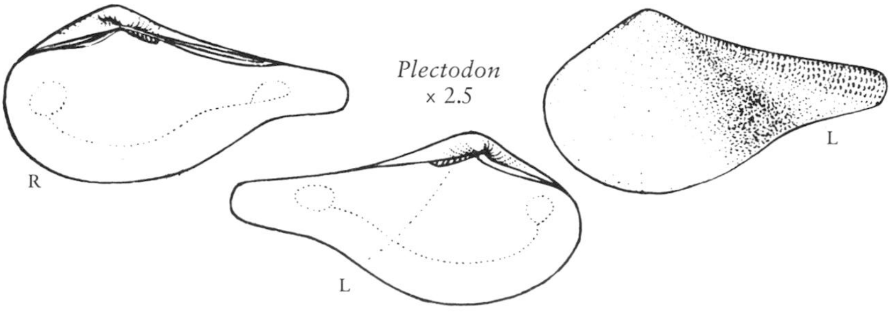

Plectodon
× 2.5

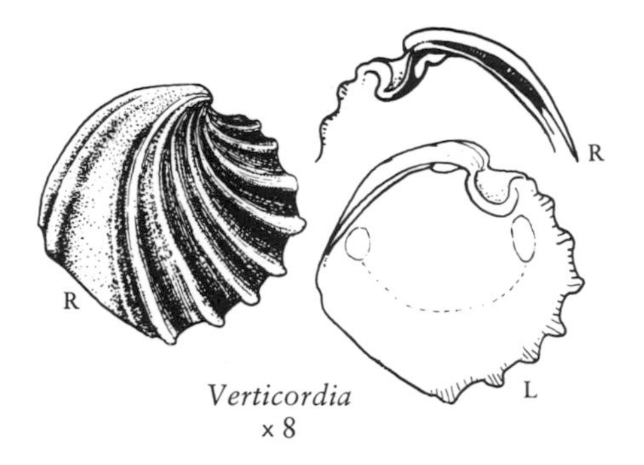

Verticordia
× 8

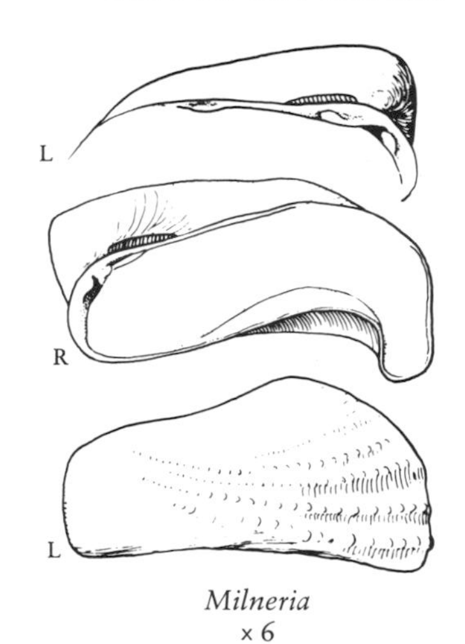

Milneria
× 6

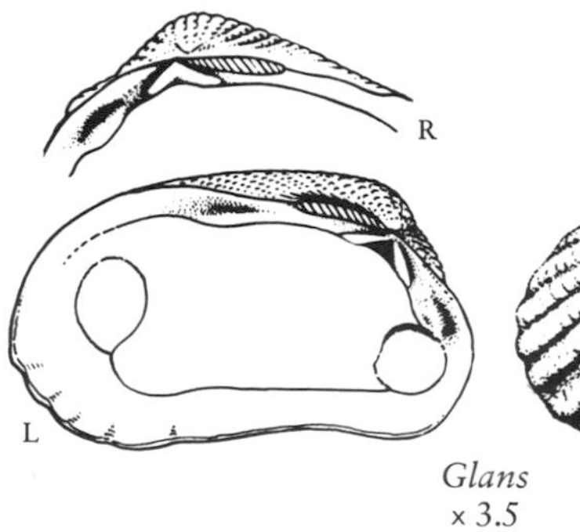

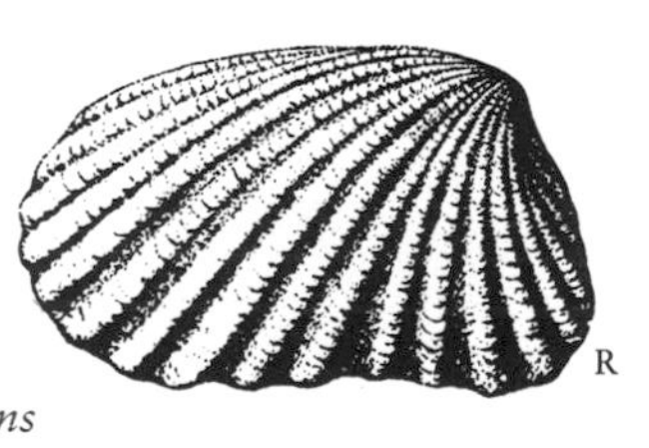

Glans
× 3.5

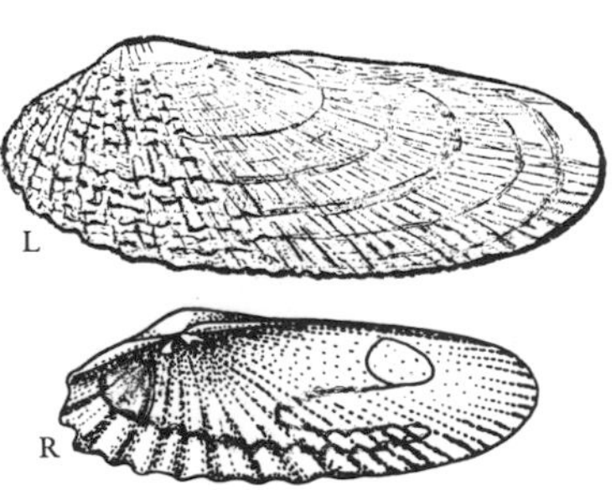

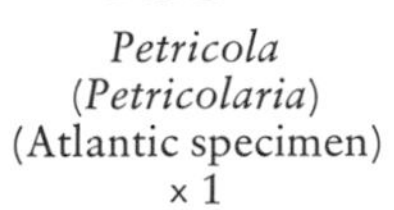

Petricola
(*Petricolaria*)
(Atlantic specimen)
× 1

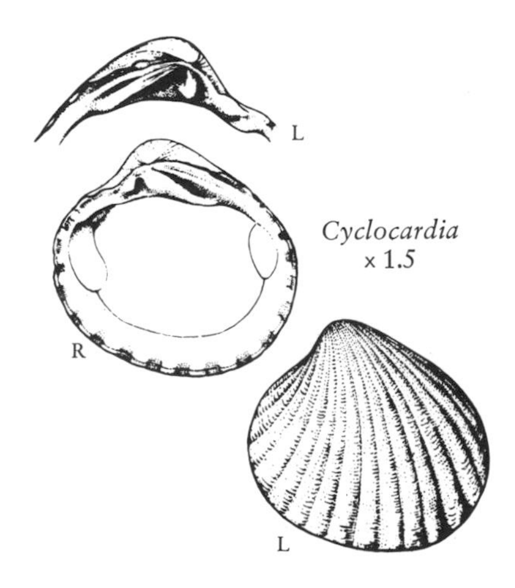

Cyclocardia
× 1.5

83(80) Sculpture of heavy radial ribs over most of the surface, showing as crenulations within (reduced in some to internal crenulations only) 84
Sculpture, if radial, only weakly expressed, mainly concentric, reticulate, or wanting 95
84(83) Interior nacreous or pearly . . . . *Verticordia*
Interior not nacreous 85
85(84) Ventral margin folded inward (in female only) . . . . *Milneria*
Ventral margin not folded inward, evenly arched 86
86(85) Quadrate; length greater than height; beaks near anterior end. . . 87
Oblique-ovate; not markedly longer than high; beaks nearly central in position 88
87(86) With lateral teeth; pallial line entire . . . . *Glans*
Without lateral teeth; with well-marked pallial sinus. . . . . *Petricola (Petricolaria)*
88(86) Hinge without clear-cut lateral teeth; periostracum velvety . . . . *Cyclocardia*
Hinge with both cardinal and lateral teeth; periostracum not evident 89
89(88) Shell small (less than 5 mm long); posterior lateral teeth wanting . . . . *Miodontiscus*
Shell medium-sized to large; posterior lateral teeth well developed 90
90(89) Posterior margin smooth or only faintly wavy 91
Posterior margin with fine to coarse crenulations 93
91(90) Ribbing well developed; beaks prosogyrate . . . . *Clinocardium*
Ribbing weak, showing as incised lines or as crenulations of the inner ventral margin; beaks orthogyrate 92

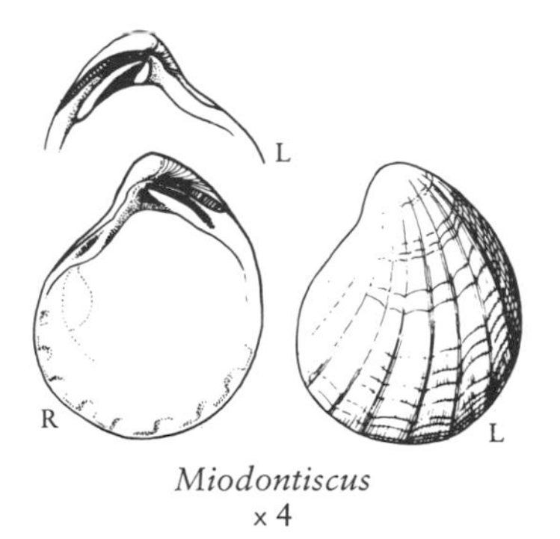

Miodontiscus
x 4

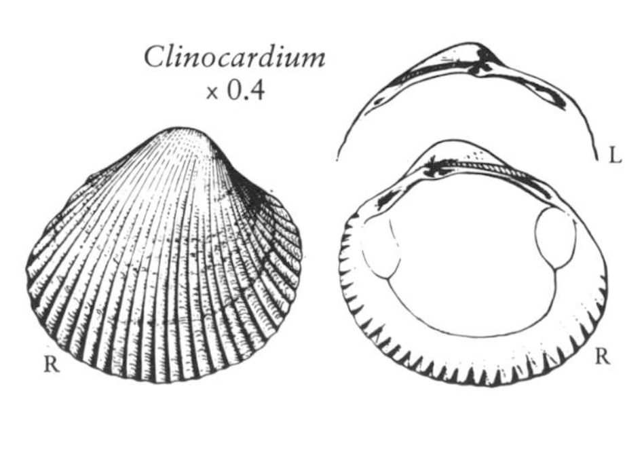

Clinocardium
x 0.4

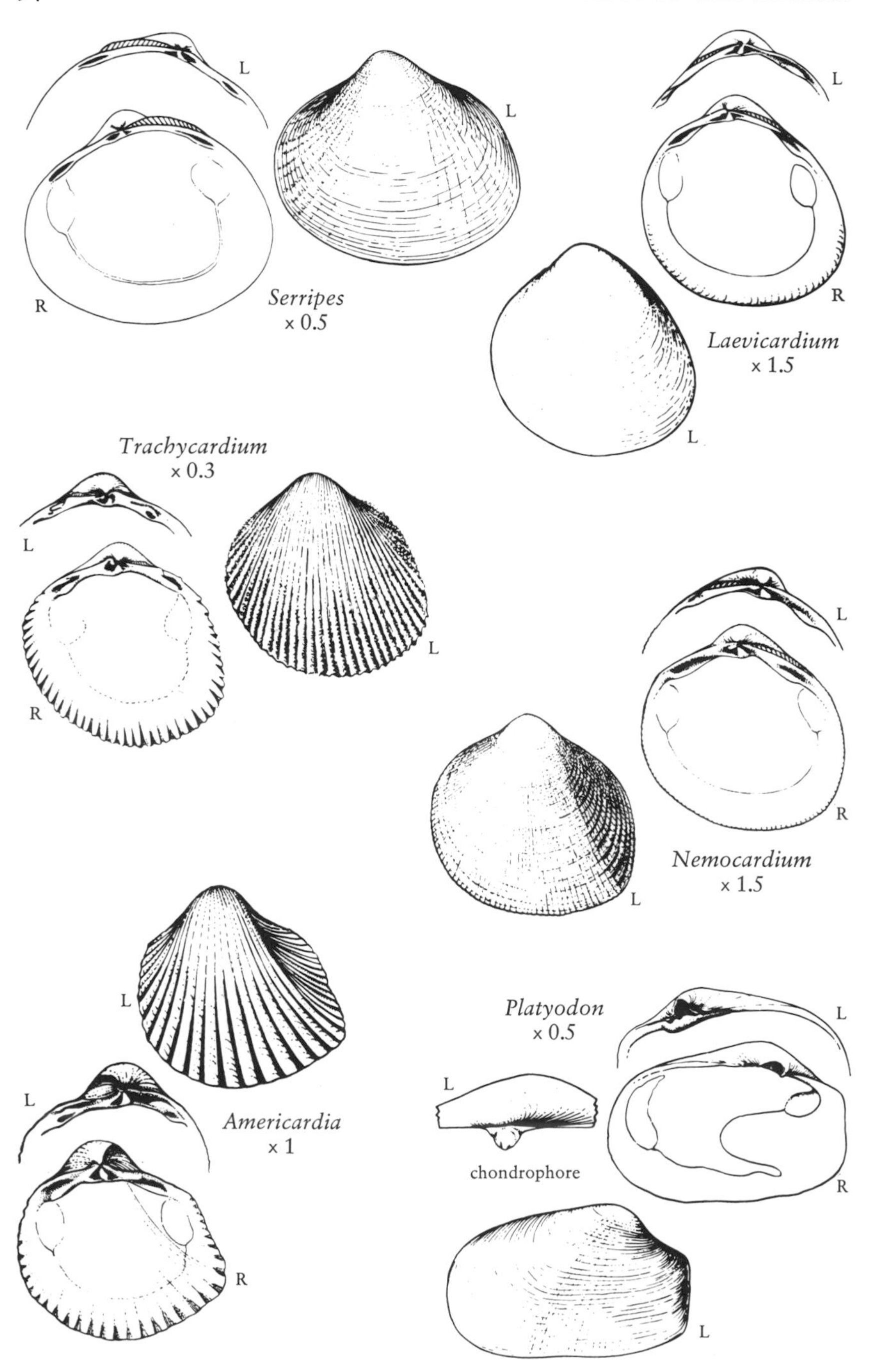
L
L
R
Serripes
× 0.5
L
R
Laevicardium
× 1.5
L
Trachycardium
× 0.3
L
L
R
L
R
Nemocardium
× 1.5
L
L
Platyodon
× 0.5
L
L
chondrophore
R
L
Americardia
× 1
R
L

92(91) Ovate-oblique; hinge teeth persistent in adult ..... *Laevicardium**
Rounded-quadrate; hinge teeth weak to obsolete in adult.... .... *Serripes**
93(90) Posterior slope of shell spinose .... *Trachycardium*
Posterior slope of shell not spinose 94
94(93) Posterior margin finely crenulate; ribs of posterior slope with overlying concentric threads .... *Nemocardium*
Posterior margin coarsely crenulate; all ribs smooth ... *Americardia*
95(83) Hinge with a chondrophore or a large, mostly horizontally projecting tooth in one valve, a socket in the other 96
Hinge either with chondrophores in both valves or with projecting teeth vertical, not horizontal 100
96(95) With a well-developed deep pallial sinus 97
Pallial line entire or nearly so 99
97(96) Concentric growth striae raised, almost lamellar ...... *Platyodon*
Concentric growth striae low, inconspicuous 98
98(97) Medium-sized to large; shell sturdy .... *Mya*
Small; shell thin .... *Sphenia*

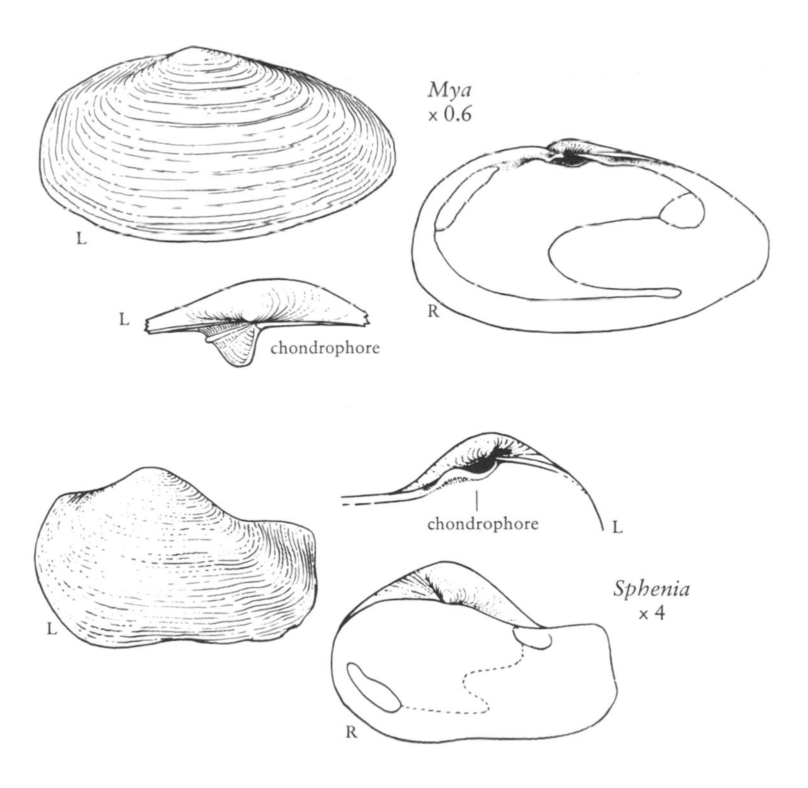

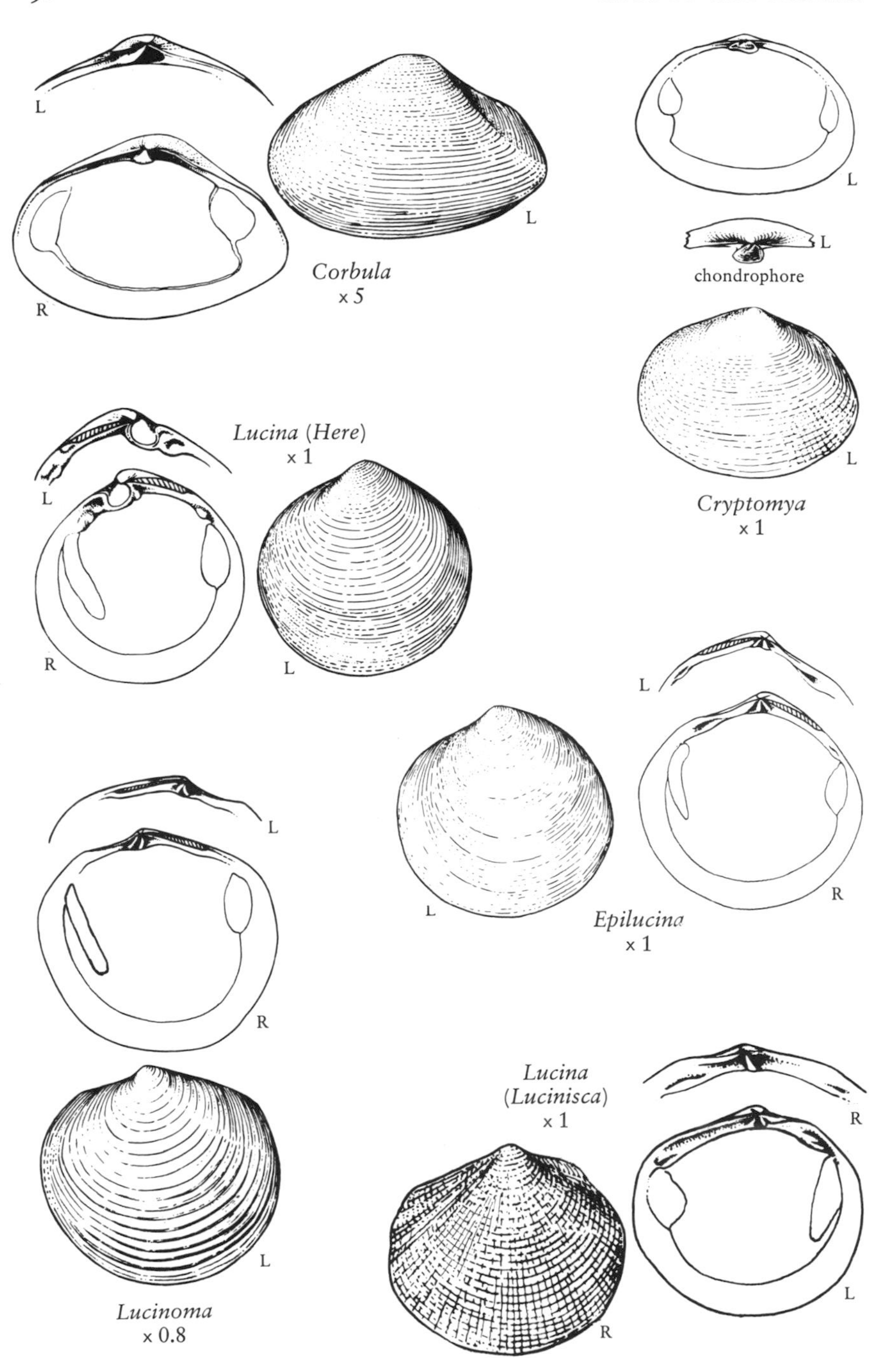

Corbula x 5

Cryptomya x 1

Lucina (*Here*) x 1

Epilucina x 1

Lucina (*Lucinisca*) x 1

Lucinoma x 0.8

99(96) Inequivalve, shell heavy for its size, with irregular concentric ribs . . . . *Corbula*
Equivalve, rather thin, smooth or with a few radial striae . . . . *Cryptomya*

100(95) Anterior adductor scar narrower than posterior, its lower end detached from pallial line and bent inward 101
Adductor scars approximately equal in shape, neither detached 105

101(100) Lunule profoundly impressed, distorting hinge line . . . *Lucina (Here)*
Lunule, if present, not impressed 102

102(101) Sculpture entirely concentric 103
Sculpture both concentric and radial 104

103(102) Concentric ribs evenly distributed, of uniform size . . . . . *Epilucina*
Ribs not evenly distributed, larger ribs widely spaced . . . . *Lucinoma*

104(102) Cancellate sculpture harsh; dorsal areas set off by a change in ribbing . . . . *Lucina (Lucinisca)*
Cancellate sculpture subdued; dorsal areas not well defined . . . . *Parvilucina*

105(100) Interior nacreous or pearly 106
Interior porcelaneous 109

106(105) Resilifer low or inconspicuous, not spoon-shaped 107
Resilifer projecting below beaks, spoon-shaped 108

107(106) Shell compressed, hinge with interlocking ridges below . . . *Pandora*
Shell inflated, hinge with a tooth in one or both valves . . . *Poromya*

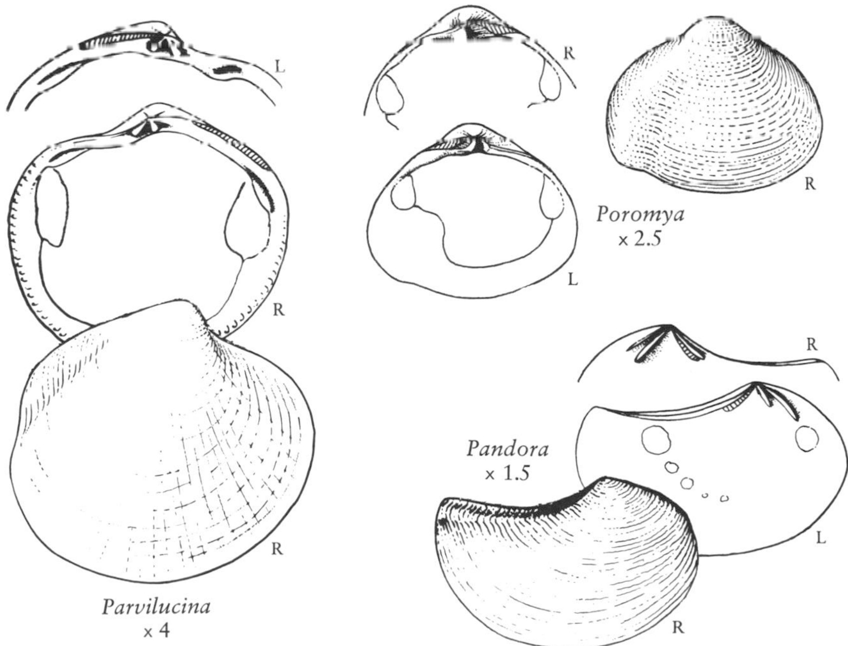

Parvilucina x 4

Poromya x 2.5

Pandora x 1.5

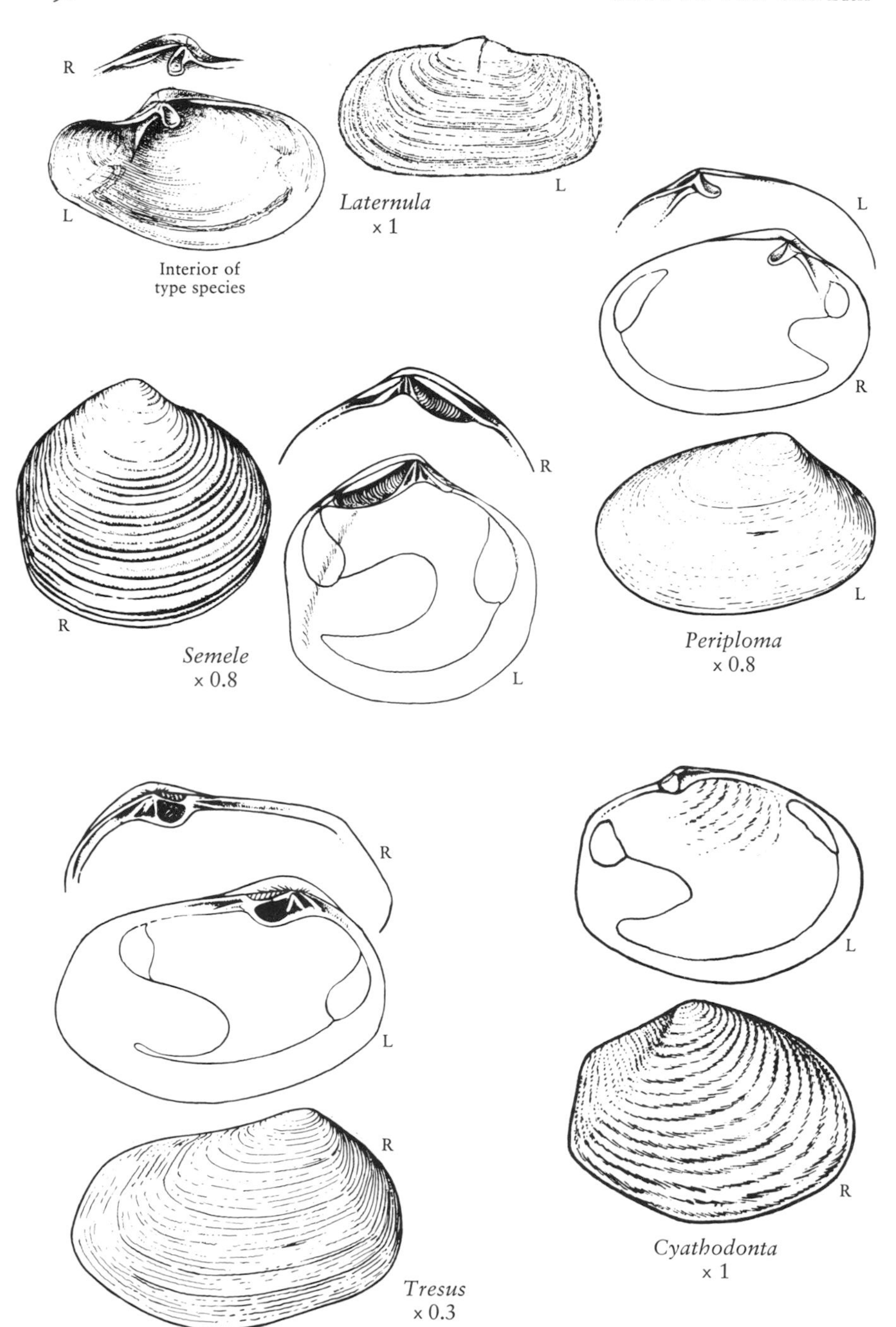
R
L
Interior of type species
Laternula
x 1
L
L
R
R
Semele
x 0.8
L
L
Periploma
x 0.8
R
L
L
R
Tresus
x 0.3
R
Cyathodonta
x 1

108(106) Posterior end produced, truncate; chondrophore not attached to a buttress . . . . *Laternula*

Posterior end not produced; chondrophore attached to a buttress . . . . *Periploma*

109(105) Ligament partly or entirely sunken or completely internal 110

Ligament external, seated on a nymph 132

110(109) With conspicuously developed pallial sinus 111

Pallial line entire or obscure 117

111(110) Ligament in a shallow resilifer that cuts across hinge plate diagonally . . . . *Semele*

Resilifer variously shaped, not cutting across hinge plate diagonally 112

112(111) Shell broadly gaping behind . . . . *Tresus*

Shell narrowly gaping or closed behind 113

113(112) With undulating concentric ribs corrugating the entire shell and showing interiorly . . . . *Cyathodonta*

Undulating sculpture, if present, confined to beak area; shell mostly smooth 114

114(113) Cardinal teeth fused, somewhat elevated, ∧-shaped; shell nearly smooth 115

Cardinal teeth weak, not fused or ∧-shaped 116

115(114) External ligament separated from internal (resilium) by a thin shelly plate or lamina . . . . *Mactra*

Ligament and resilium not separated by a lamina, though structurally distinct . . . . *Spisula*

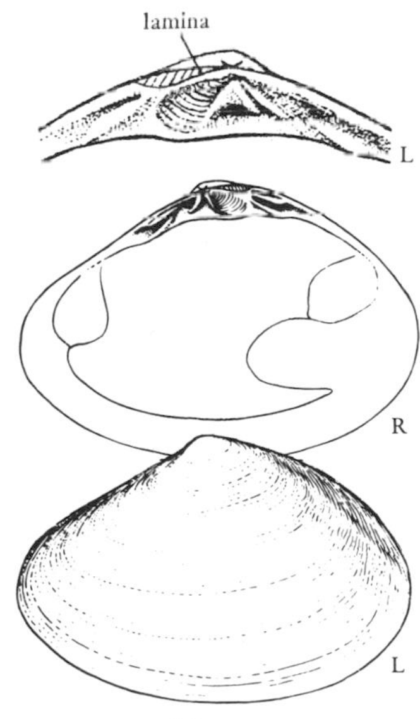

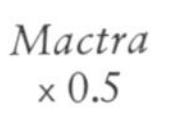
Mactra
× 0.5

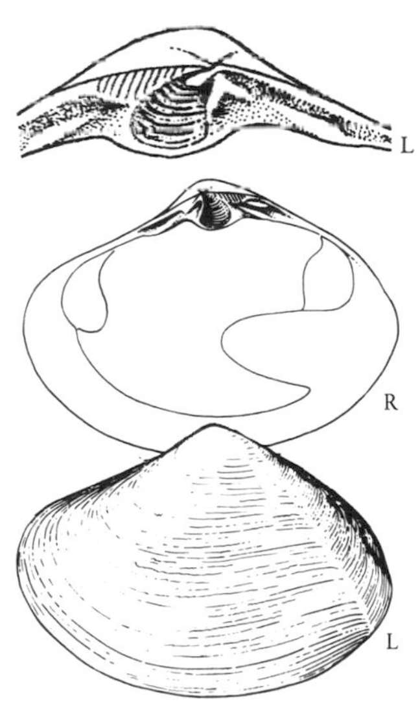

Spisula
× 0.5

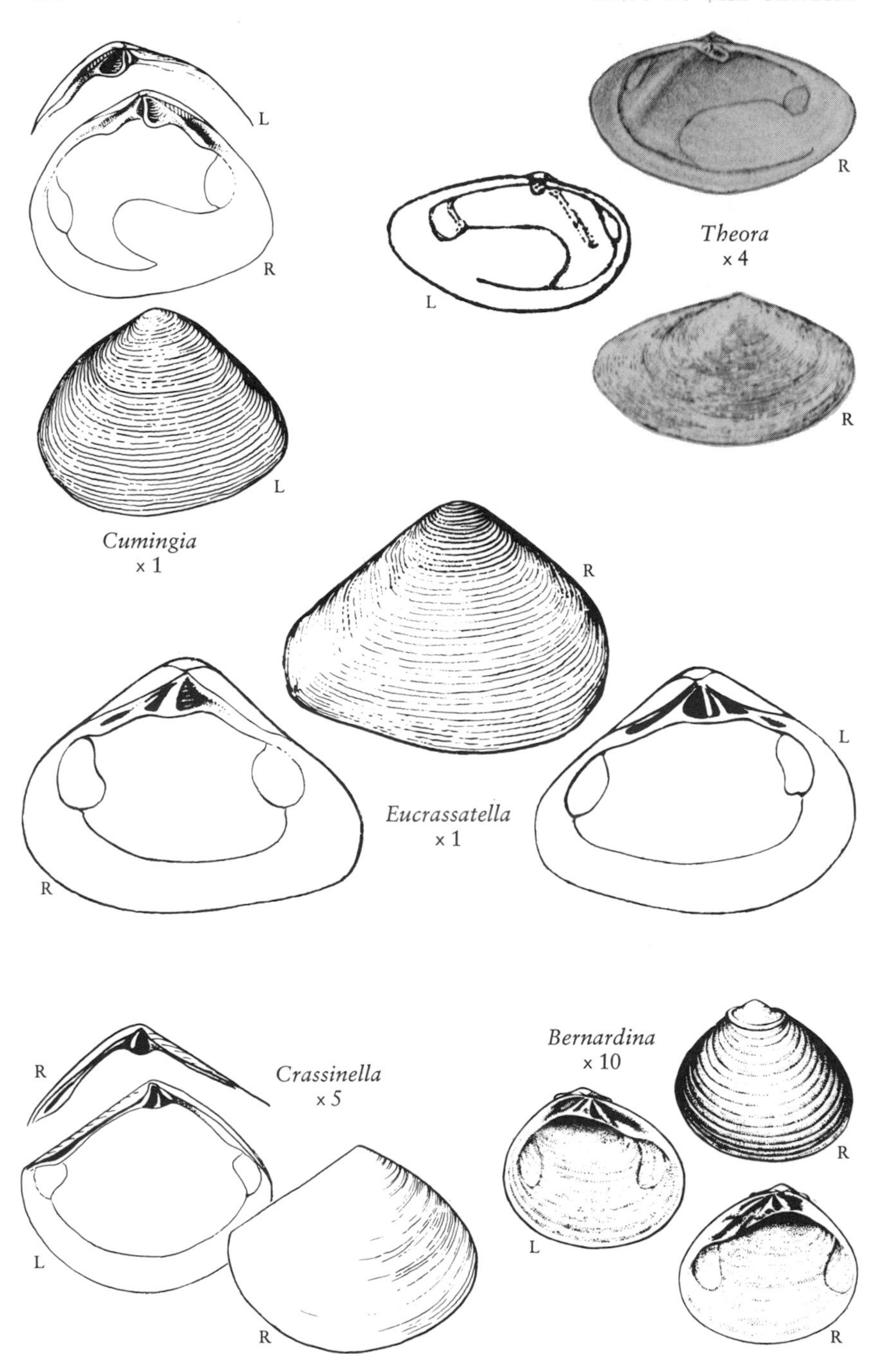

Theora × 4

Cumingia × 1

Eucrassatella × 1

Crassinella × 5

Bernardina × 10

116(114) Shell with lamellar concentric sculpture .... *Cumingia*
Shell smooth, thin .... *Theora*
117(110) Sculpture of concentric ribs, at least on umbones 118
Shell smooth or nearly so, no concentric ribs 120
118(117) Sculpture confined to umbonal area; shell heavy for its size....
.... *Eucrassatella*
Sculpture uniform over entire shell; shell thin, small 119
119(118) Outline trigonal .... *Crassinella*
Outline ovate .... *Bernardina*
120(117) Markedly inequilateral; beaks nearly at anterior end 121
Moderately to only slightly inequilateral 122
121(120) Cardinal teeth wanting in one valve, smooth .... *Mysella*
Cardinal teeth two in each valve, serrate .... *Pristes*
122(120) Hinge with one or more lateral teeth 123
Hinge with cardinal teeth only, no laterals 129
123(122) Adults more than 10 mm in length; with greenish periostracum
.... *Kellia*
Adults less than 10 mm in length; periostracum, if present, not greenish 124

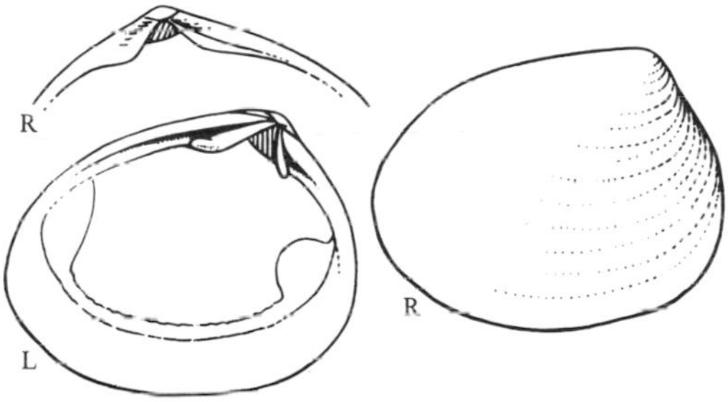

R R L *Mysella* × 10

Kellia × 1.5 R R L

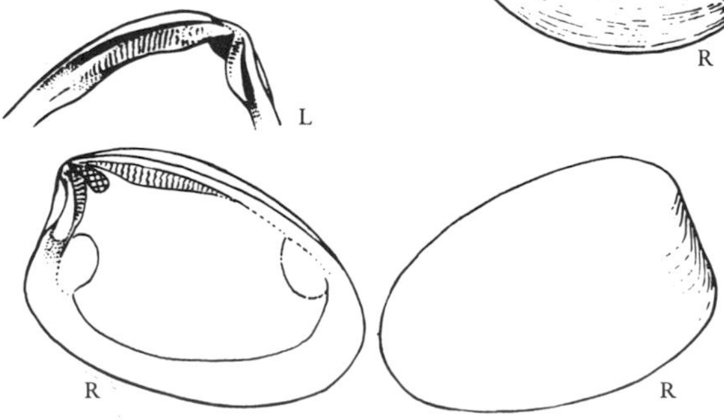

L R R *Pristes* × 8

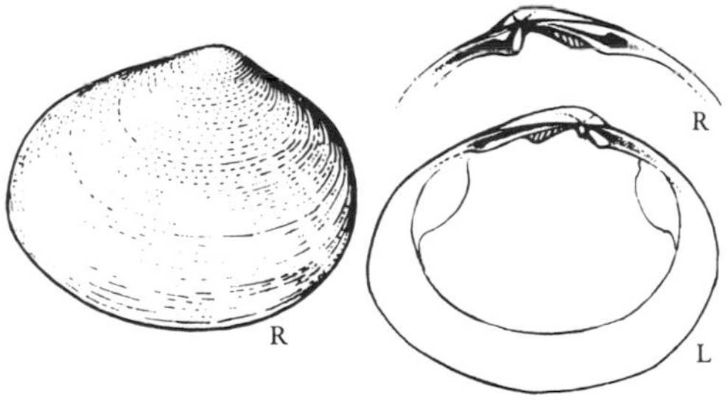

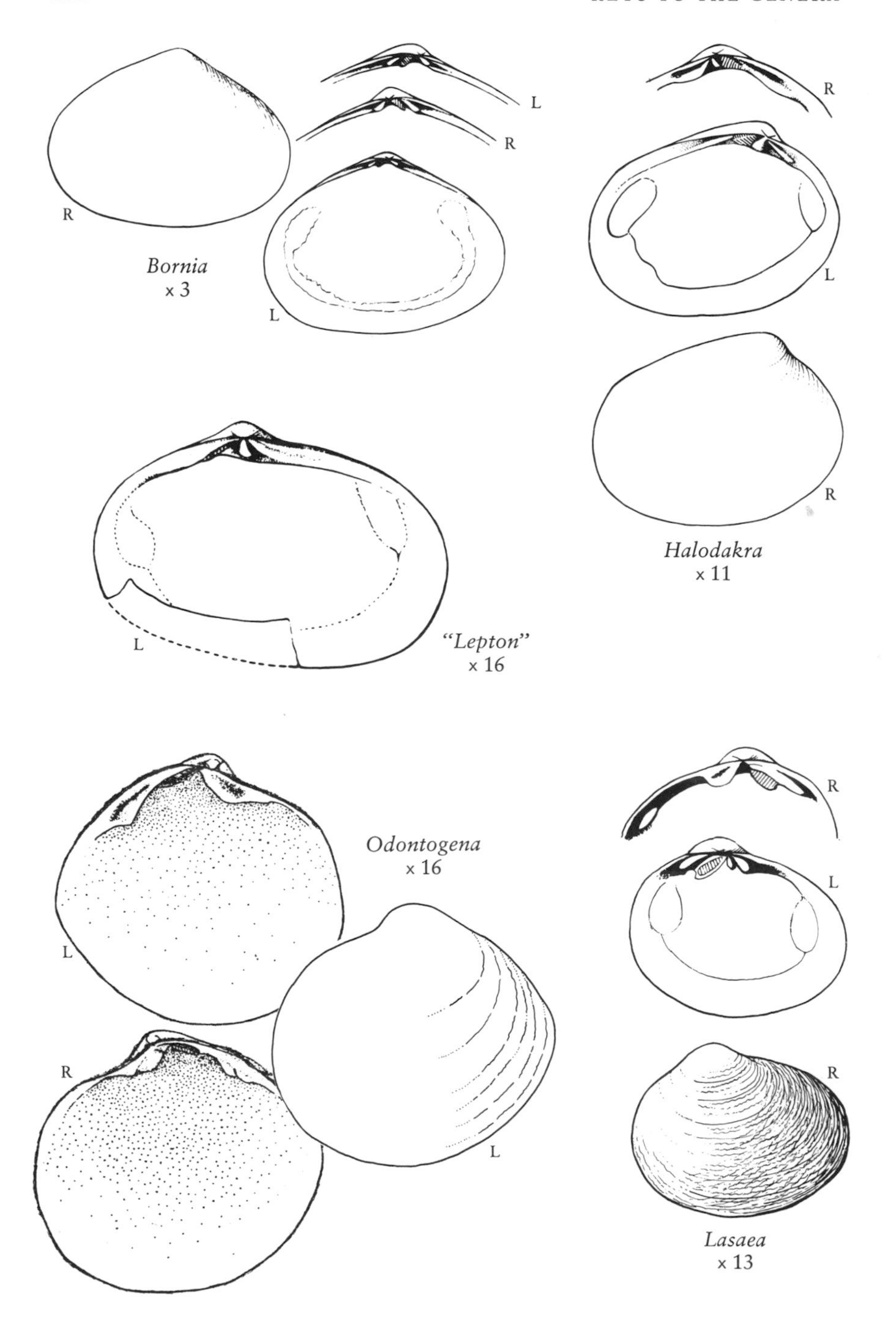

Bornia x 3

Halodakra x 11

"Lepton" x 16

Odontogena x 16

Lasaea x 13

124(123) Nearly trigonal; surface with microscopically fine irregular radial striae or punctations . . . . *Bornia*
Somewhat inequilateral; surface smooth or with concentric striae 125
125(124) With at least two radiating hinge teeth anterior to resilifer 126
With one cardinal tooth in each valve 127
126(125) Ovate; hinge plate well developed . . . . *Halodakra*
Elliptical; hinge plate narrow, indented . . . . *"Lepton"*
127(125) Periostracum wrinkled; cardinal tooth thornlike . . . . *Lasaea*
Periostracum smooth or wanting; cardinal tooth long or heavy 128
128(127) Hinge teeth large but poorly defined, projecting below hinge plate . . . . *Odontogena*
Hinge teeth elongate, not projecting below hinge plate . . . . *Montacuta*
129(122) Outline suborbicular . . . . *Tomburchus**
Outline not suborbicular 130
130(129) Trigonal; beaks high, pointed . . . .*Aligena*
Quadrate; beaks low, not pointed 131

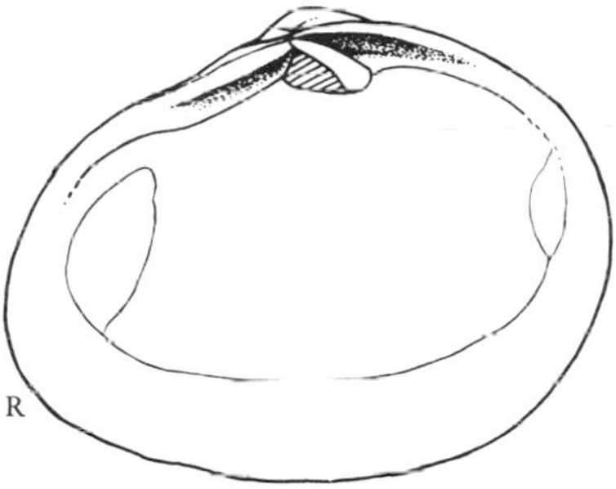

R

Montacuta
x 15

R

L

R

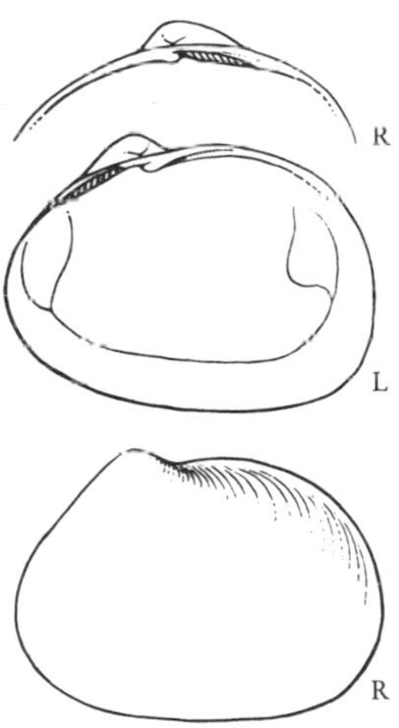

Tomburchus
x 12

L

L

R

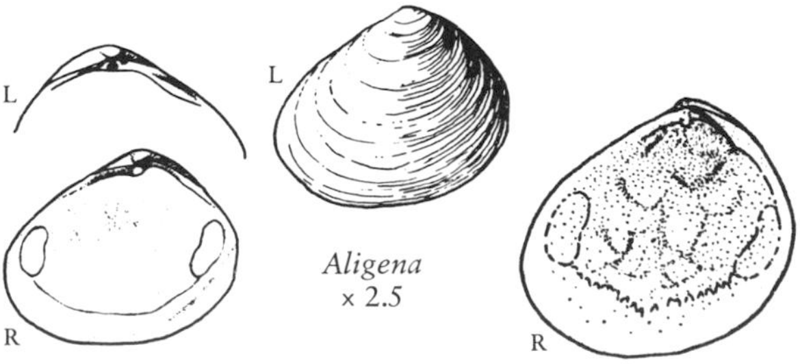

Aligena
x 2.5

R

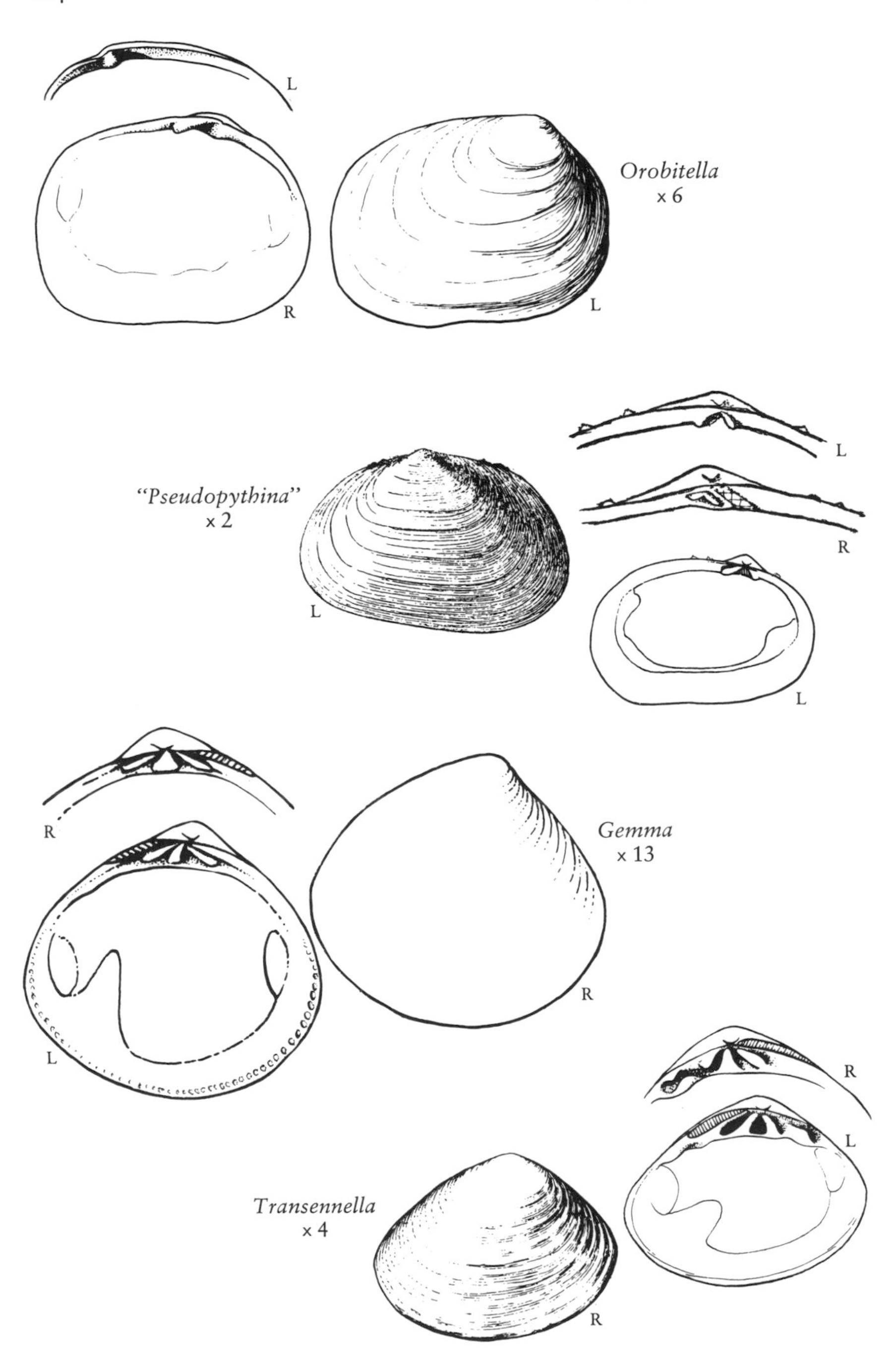
L
R
L
Orobitella
x 6
"Pseudopythina"
x 2
L
R
L
L
R
Gemma
x 13
L
R
R
L
Transennella
x 4
R

131(130) Right valve with projecting hinge tooth; valves often with shaggy periostracum .... *Orobitella*
Right valve with short hinge teeth; no periostracum but dorsal margin with short spines or prickles, especially in adults... .... *"Pseudopythina"*
132(109) Adults small to minute (less than 10 mm in length) 133
Adults more than 10 mm in length 139
133(132) Hinge with three cardinal teeth in each valve 134
Hinge not having three cardinal teeth in both valves 136
134(133) Pallial sinus bent sharply upward; inner ventral margin finely crenulate .... *Gemma*
Pallial sinus rounded; inner margin not crenulate 135
135(134) Hinge with conspicuous anterior lateral teeth; a few oblique grooves on inner ventral margin .... *Transennella*
Hinge without lateral teeth; inner margin smooth ...... *Psephidia*
136(133) Cardinal teeth wanting in left valve .... *Grippina*
At least one cardinal tooth in each valve 137
137(136) Suborbicular; area in front of beaks flattened .... *Axinopsida*
Ovate; area in front of beaks not flattened 138

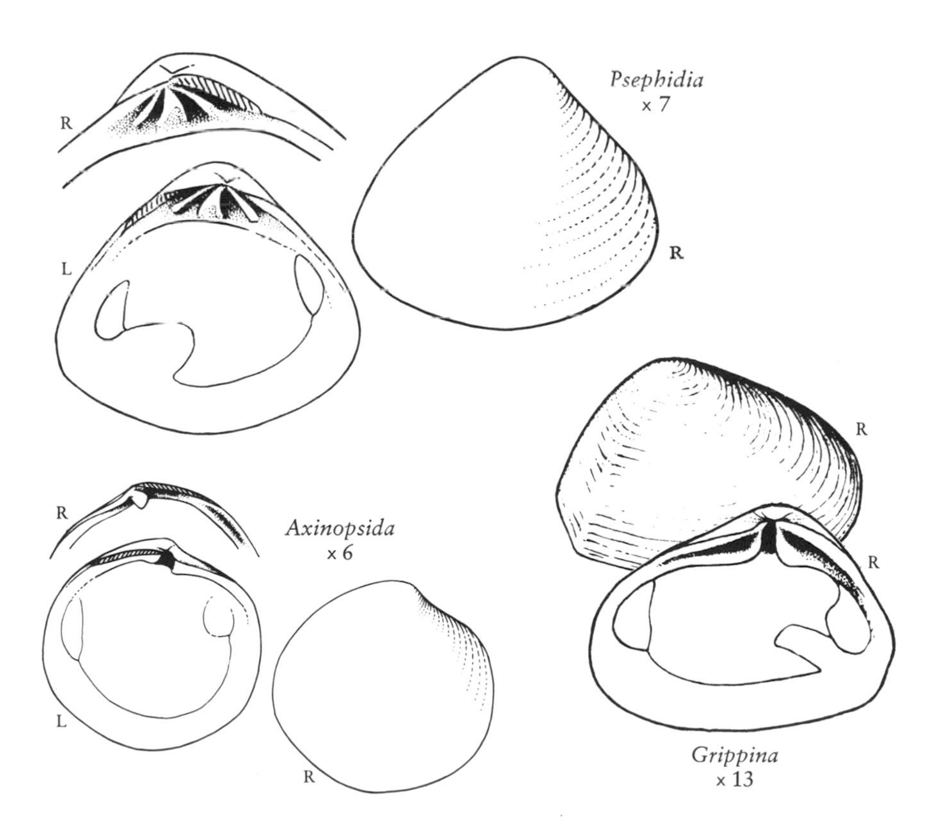

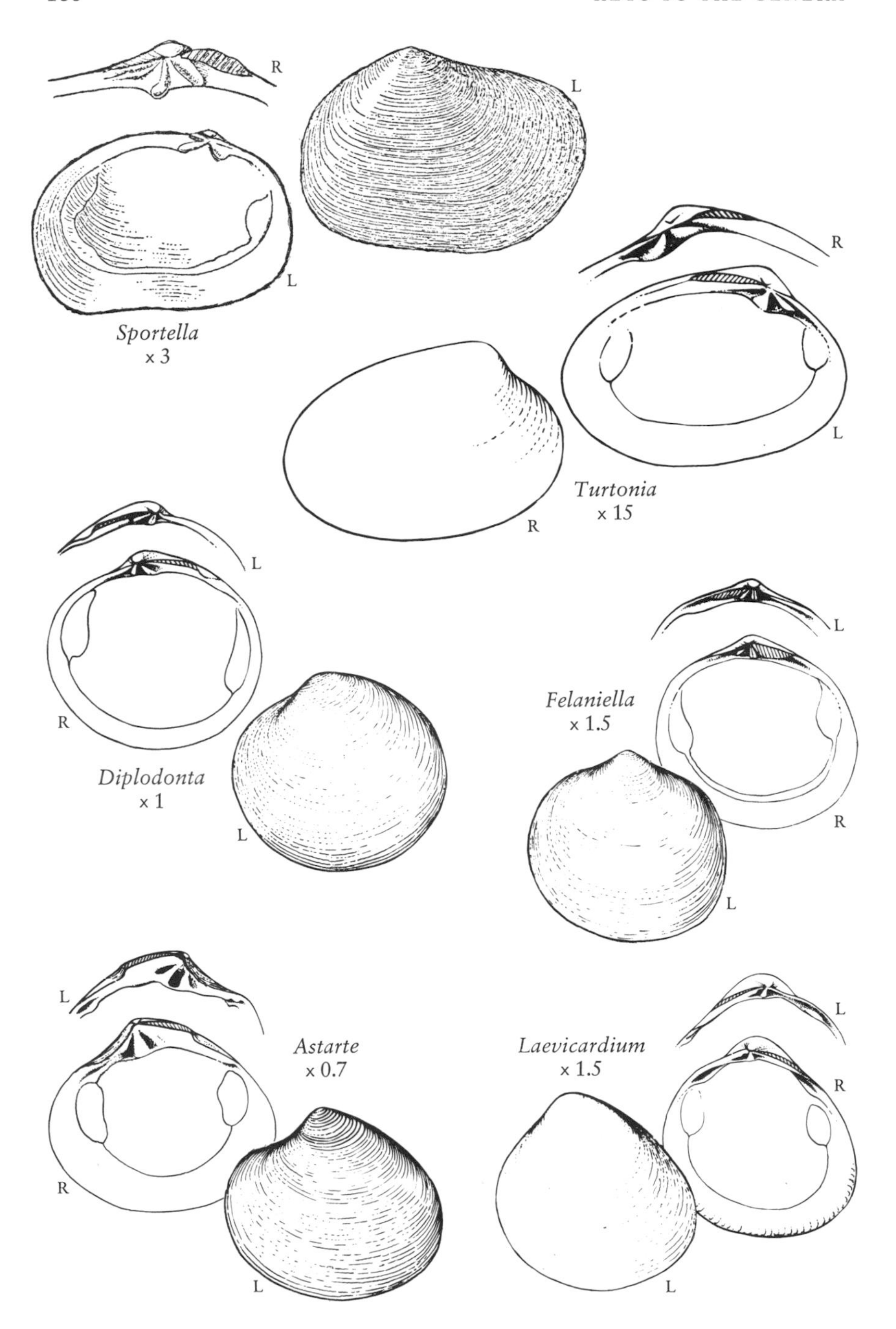

Sportella x 3

Turtonia x 15

Diplodonta x 1

Felaniella x 1.5

Astarte x 0.7

Laevicardium x 1.5

138(137) Light-colored; mantle scar wide, irregular .... *Sportella*
Dark-colored; mantle scar narrow, even .... *Turtonia*
139(132) Pallial line continuous and smoothly arched, without a sinus ...140
Pallial line somewhat broken, either with a sinus or present only in patches 145
140(139) Outline subcircular 141
Outline ovate-trigonal to elongate 142
141(140) Inflated; periostracum papery in texture .... *Diplodonta*
Lens-shaped; periostracum shiny, adherent .... *Felaniella*
142(140) Trigonal to ovate, with sculpture (at least with crenulations on ventral margin) 143
Elongate, unsculptured 144
143(142) Shell sturdy, sculpture concentric, present at least on beaks and umbones .... *Astarte*
Shell thin; no concentric sculpture on beaks or umbones (except growth lines) .... *Laevicardium**
144(142) Without lateral teeth .... *Vesicomya*
With an anterior lateral tooth in each valve .... *Calyptogena*
145(139) Cardinal teeth three in each valve 146
Cardinal teeth not three in each valve 159
146(145) Lateral teeth present 147
Lateral teeth none 151

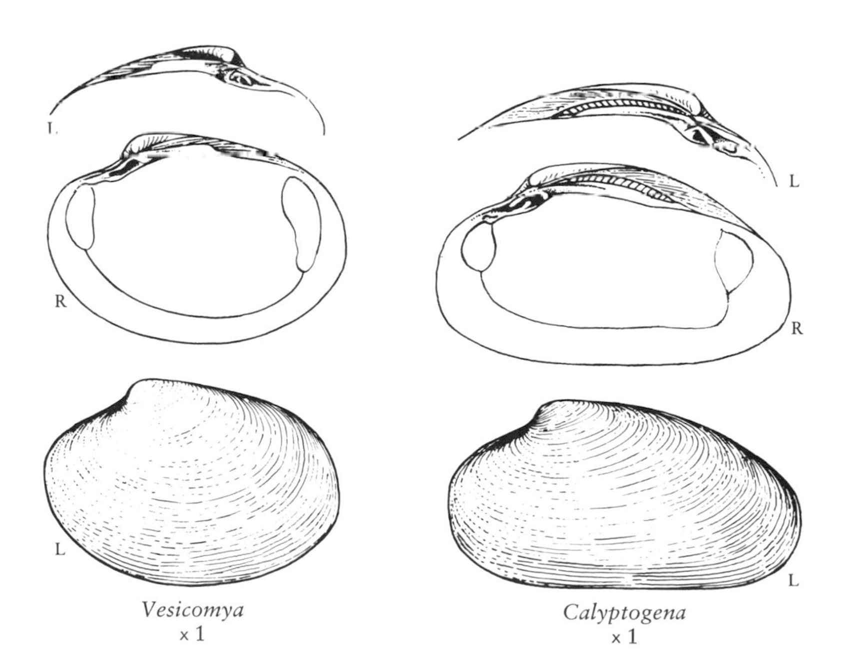

Vesicomya
x 1

Calyptogena
x 1

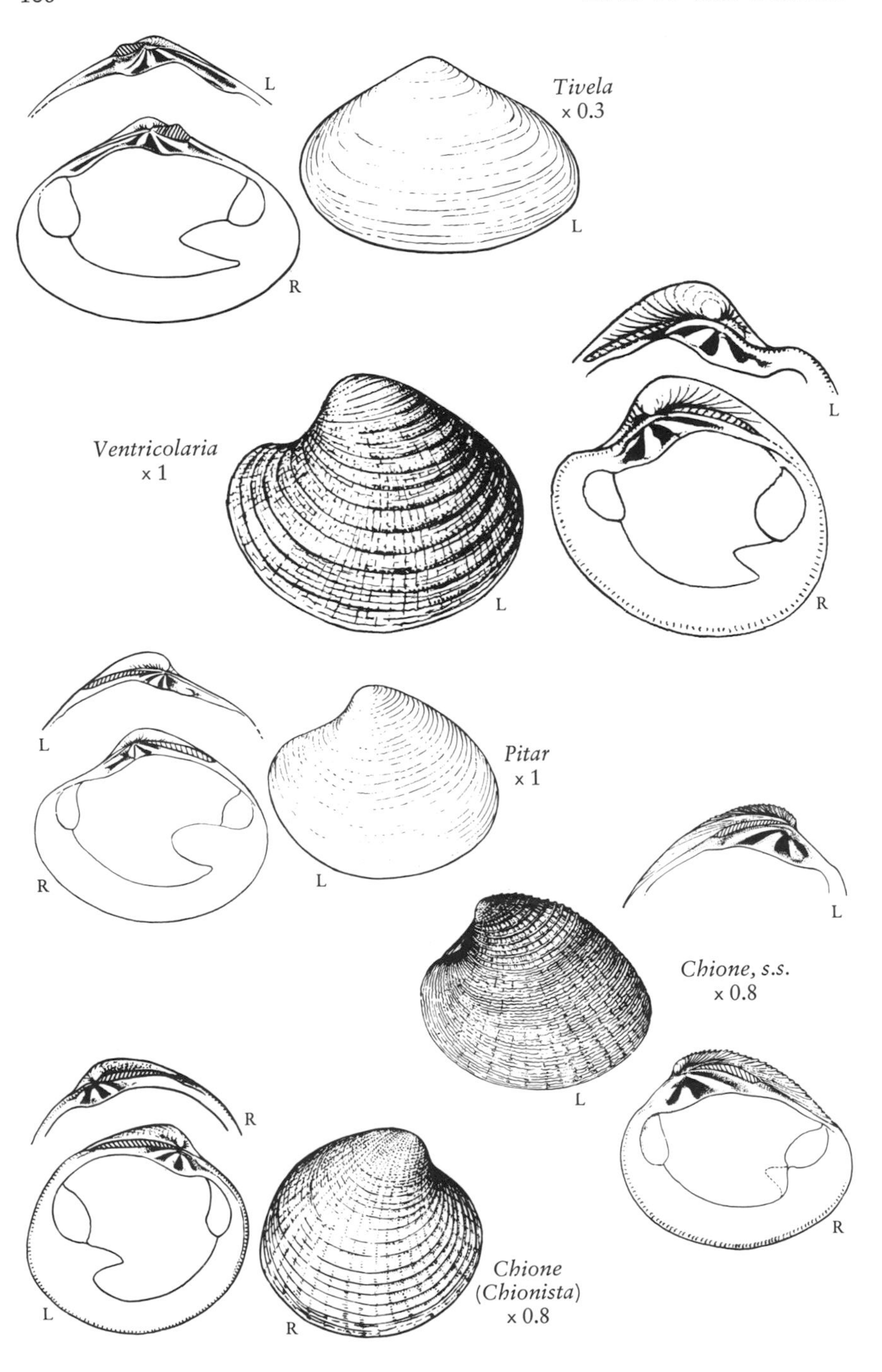
L
Tivela
x 0.3
L
R
L
Ventricolaria
x 1
L
R
L
Pitar
x 1
R
L
L
Chione, s.s.
x 0.8
L
R
R
L
Chione
(Chionista)
x 0.8
R

147(146) Outline trigonal or subtrigonal .... *Tivela*
Outline inequilateral, not trigonal 148
148(147) Radial sculpture showing in interspaces and on crenulate inner margin .... *Ventricolaria*
Radial sculpture wanting, inner margin smooth 149
149(148) Shell smooth .... *Pitar*
With well-developed concentric sculpture 150
150(149) Concentric ribs smooth, polished, sometimes dividing; posterior end rounded .... *Amiantis*
Concentric ribs rough, not polished, not dividing; posterior end truncate .... *Saxidomus*
151(146) Shell markedly sculptured 152
Shell smooth, with growth striae only 158
152(151) Pallial sinus shallow; outline short-ovate 153
Pallial sinus moderate to deep; outline elongate trigonal to quadrate 154
153(152) With a well-developed lunule and escutcheon .... *Chione, s.s.*
Lunule and escutcheon not set off .... *Chione (Chionista)*

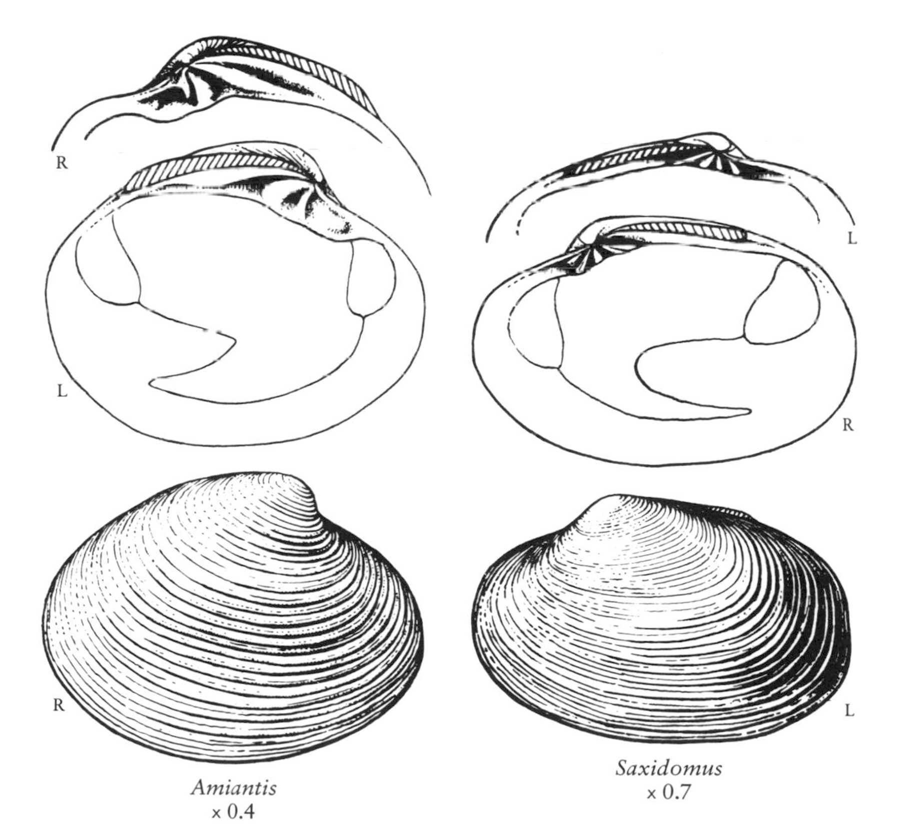

Amiantis
× 0.4

Saxidomus
× 0.7

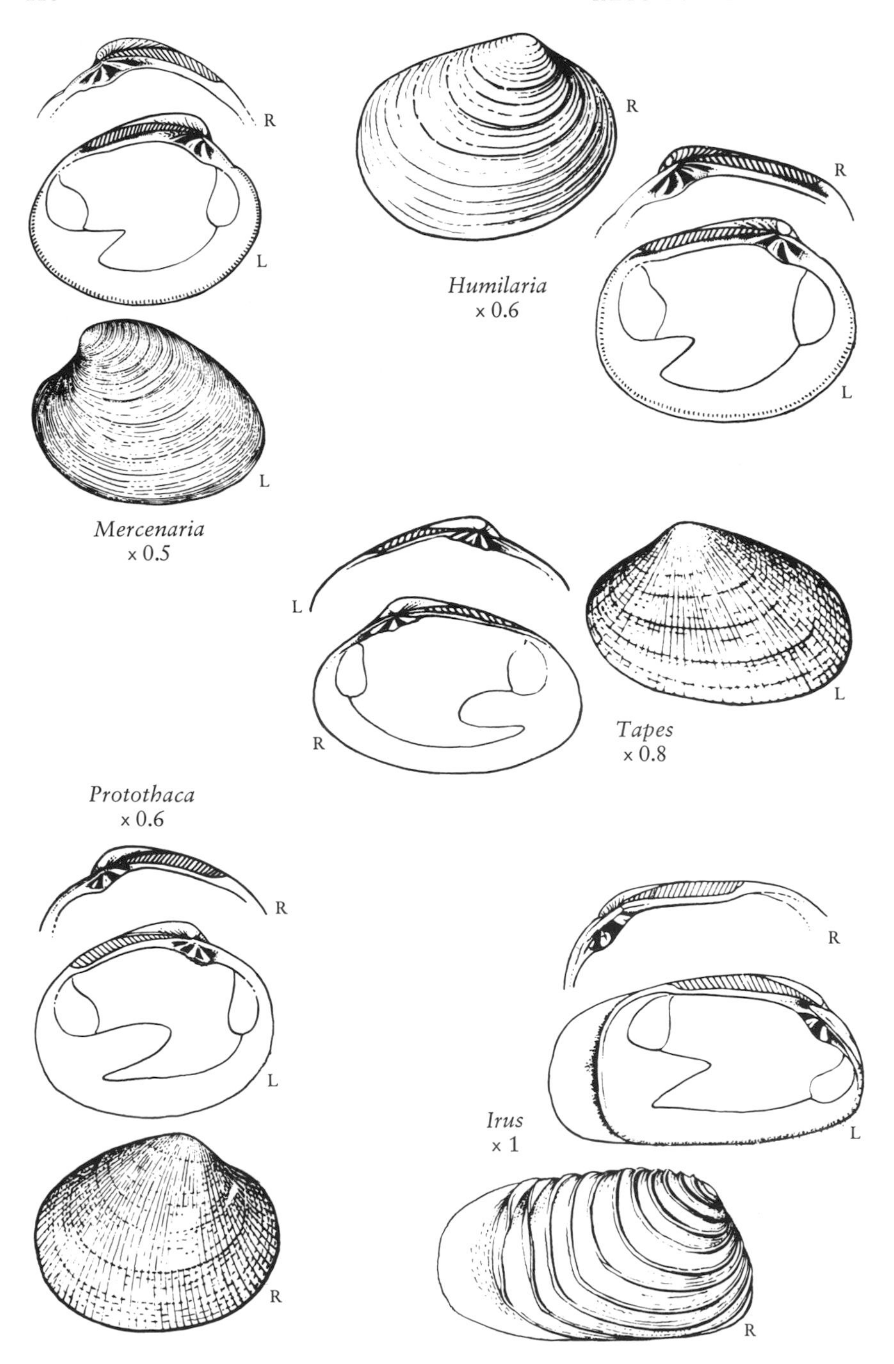

Mercenaria
× 0.5

Humilaria
× 0.6

Tapes
× 0.8

Protothaca
× 0.6

Irus
× 1

154(152) Concentric sculpture fine or closely spaced, radial sculpture wanting or as marginal crenulations only 155
Concentric sculpture either widely spaced or intersected by numerous radial riblets 156
155(154) Outline elliptical, posterior margin joining dorsal margin in an even arc or curve .... *Mercenaria*
Outline subquadrate, posterior margin meeting dorsal margin at almost a right angle .... *Humilaria*
156(154) Concentric sculpture laminar, widely spaced; radial striae only on umbones; inner margin crenulate .... *Irus*
Concentric sculpture generally low, intersected by radial riblets; inner margin mostly smooth 157
157(156) Hinge plate narrow, its ventral edge a smooth curve; pallial sinus moderately deep .... *Tapes*
Hinge plate relatively wide, its ventral edge with an angulate bend; pallial sinus very deep, rounded .... *Protothaca*
158(151) With a polished periostracum .... *Liocyma*
Without periostracum .... *Compsomyax*
159(145) Cardinal teeth three in one valve, two in other 160
Cardinal teeth fewer than three in both valves 161

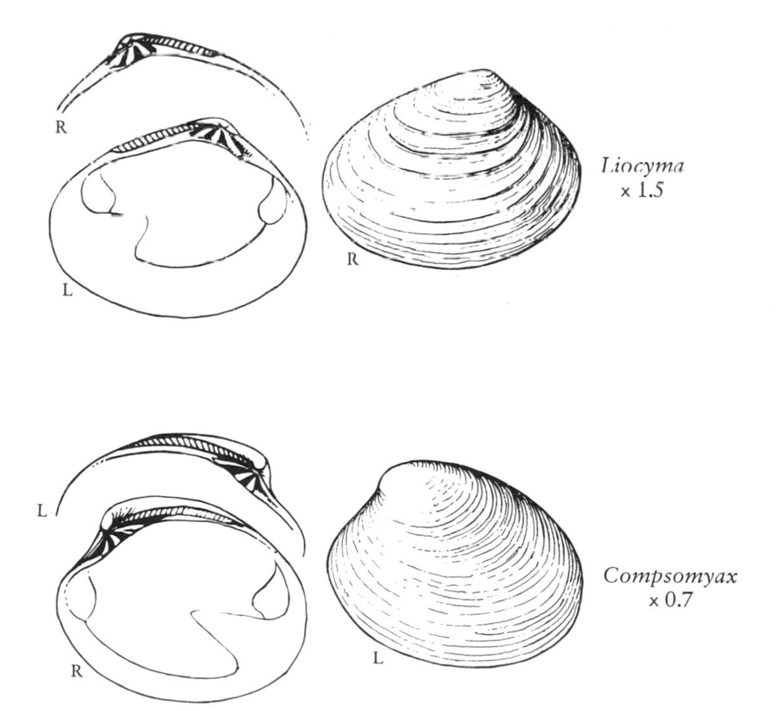

Liocyma x 1.5

Compsomyax x 0.7

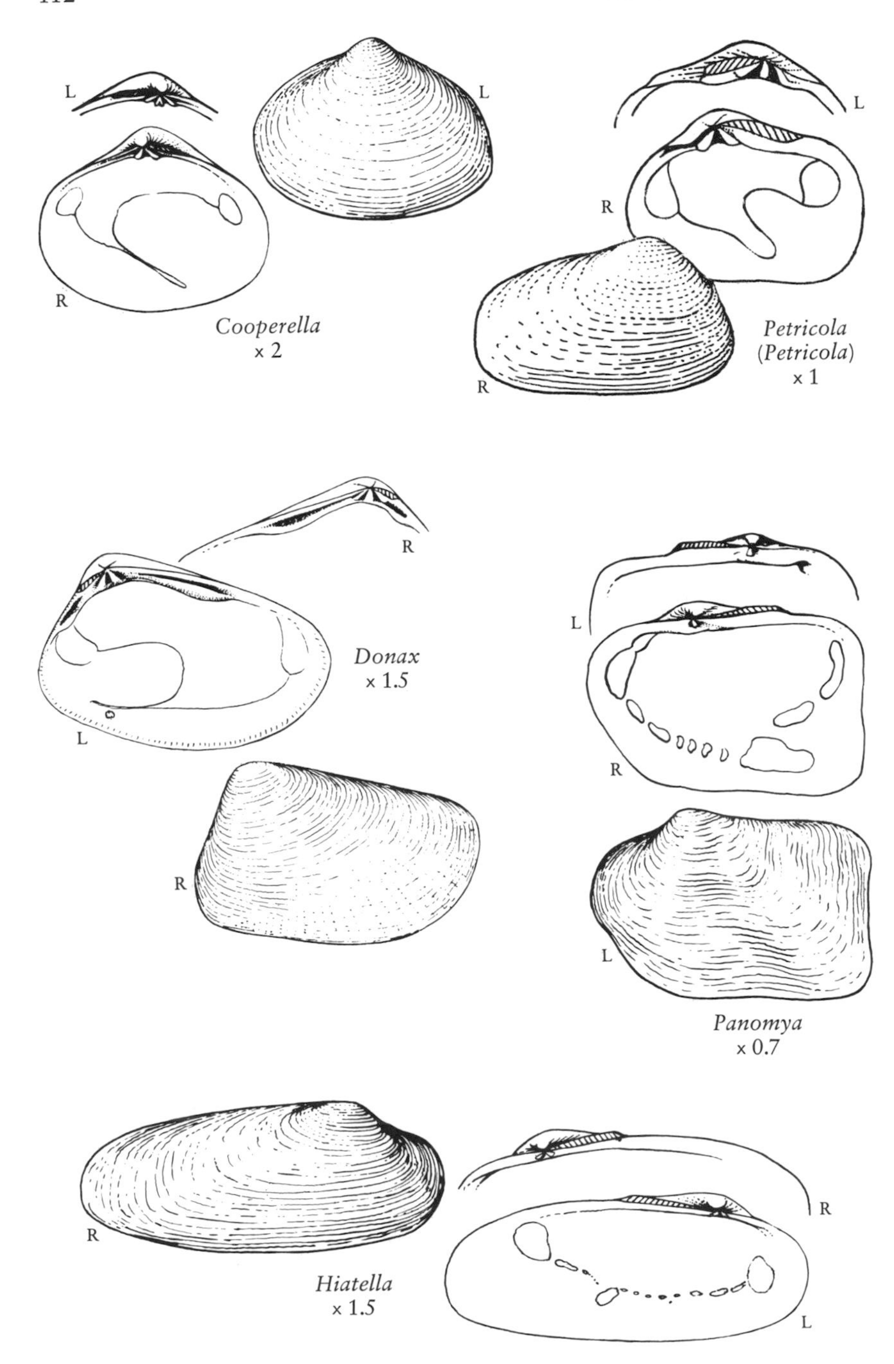

Cooperella
x 2

Petricola
(*Petricola*)
x 1

Donax
x 1.5

Panomya
x 0.7

Hiatella
x 1.5

160(159) Shell thin, polished .... *Cooperella*
Shell sturdy, of chalky texture, not polished ... *Petricola (Petricola)*
161(159) Inner margins crenulate, at least at ends .... *Donax*
Inner margins smooth throughout 162
162(161) Pallial line of discontinuous scars 163
Pallial line continuous 164
163(162) Large (over 50 mm), pallial scars conspicuous; hinge with one pointed tooth .... *Panomya*
Small to medium-sized, pallial line faint; hinge edentulous in adult, two small teeth visible in younger stages .... *Hiatella**
164(162) Pallial line broad, pallial sinus short .... *Panopea*
Pallial line narrow, pallial sinus large 165
165(164) Inequivalve, right valve flatter than left .... *Nuttallia*
Equivalve with respect to convexity (not necessarily with respect to symmetry—one or both valves may be flexed) 166

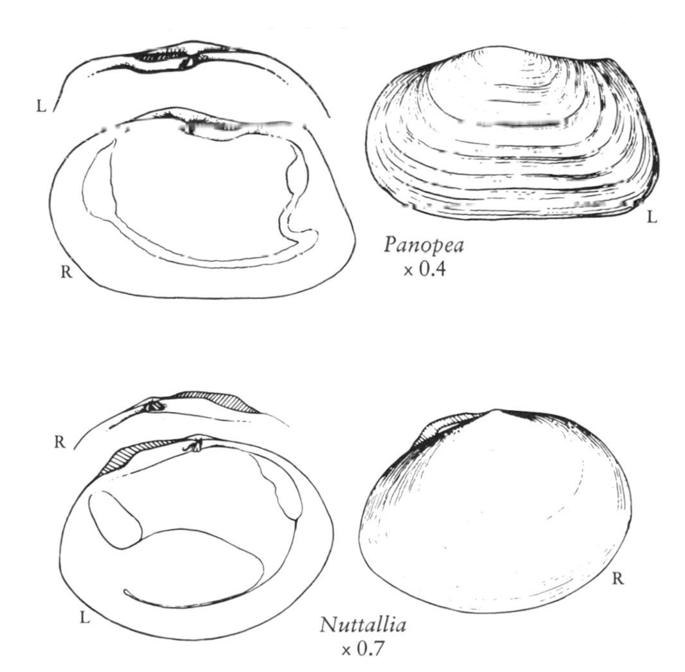

Panopea
x 0.4

Nuttallia
x 0.7

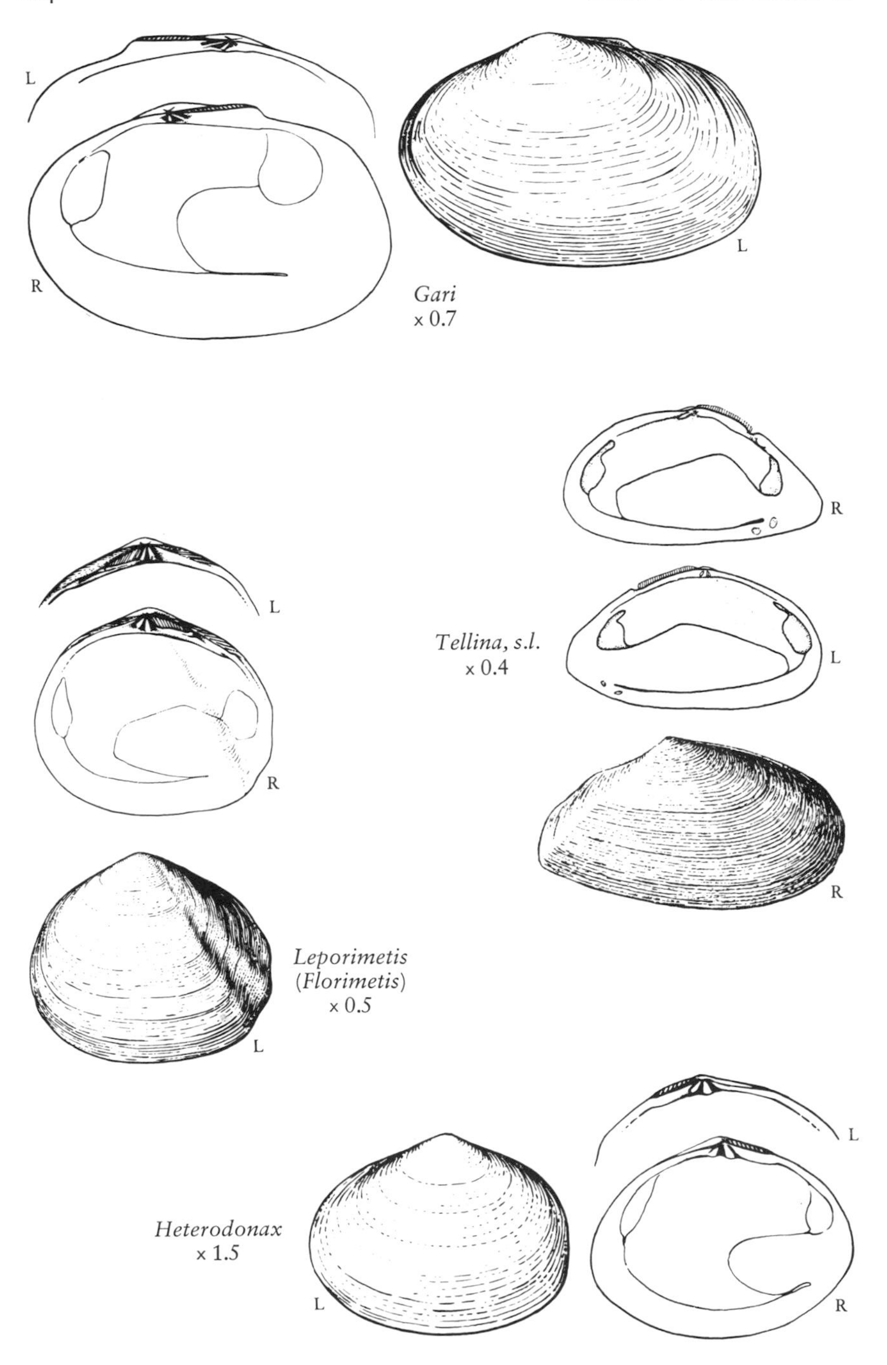

Gari
× 0.7

Tellina, s.l.
× 0.4

Leporimetis (Florimetis)
× 0.5

Heterodonax
× 1.5

166(165) Posterior end with a moderately wide gape; hinge plate wide; ligament on a broad nymph .... *Gari*
Posterior end closed or only narrowly gaping; hinge plate moderately wide to narrow; ligament not set up on a high nymph 167
167(166) Hinge with lateral teeth in one or both valves .... *Tellina, s.l.*
Hinge without lateral teeth 168
168(167) Ligament partially sunk below hinge margin; posterior slope with strong flexure; umbonal area inflated ....
.... *Leporimetis (Florimetis)*
Ligament not at all sunken below hinge margin; posterior slope without a strong flexure; umbonal area not markedly inflated 169
169(168) Pallial sinus striate; entire shell usually colored, ranging from off-white to purple or pink, with or without radial color lines externally .... *Heterodonax*
Pallial sinus not striate; color, if present, as a general internal suffusion, no external color stripes 170
170(169) Shell heavy, elongate, not flexed; interior mostly with a pink or orange (sometimes yellow) suffusion ..... *Tellina (Megangulus)*
Shell mostly thin, slightly to moderately flexed, especially in heavier or more elongate specimens; never with internal coloration .... *Macoma*

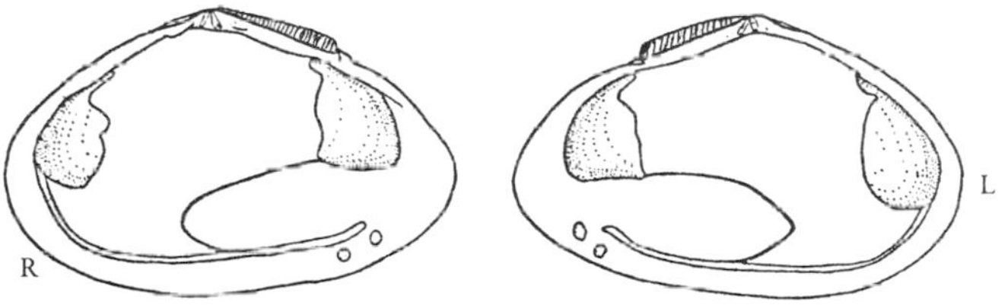

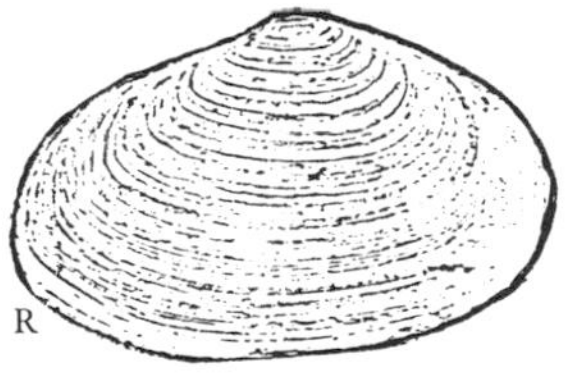

R L R

Tellina
(*Megangulus*)
x 0.5

Macoma
x 0.6

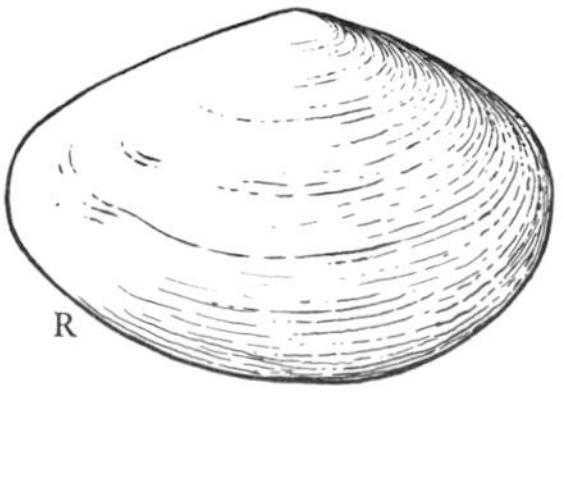

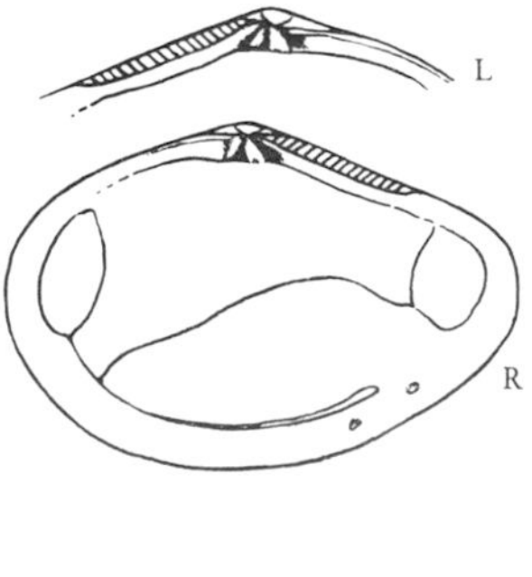

L R R

KEY TO THE POLYPLACOPHORA

All chiton specimens to be identified should have the girdle preserved. Preferably, there should be two or more specimens, one kept intact (either dried or in alcohol), the other disarticulated to reveal the margins of the separate valves. Although the sculpture and color patterning of the valves are by far the easiest features to observe, they are inadequate guides for either classification or identification, except for a few forms. Rather, the primary diagnostic characters reside in both the valve structure and the girdle.

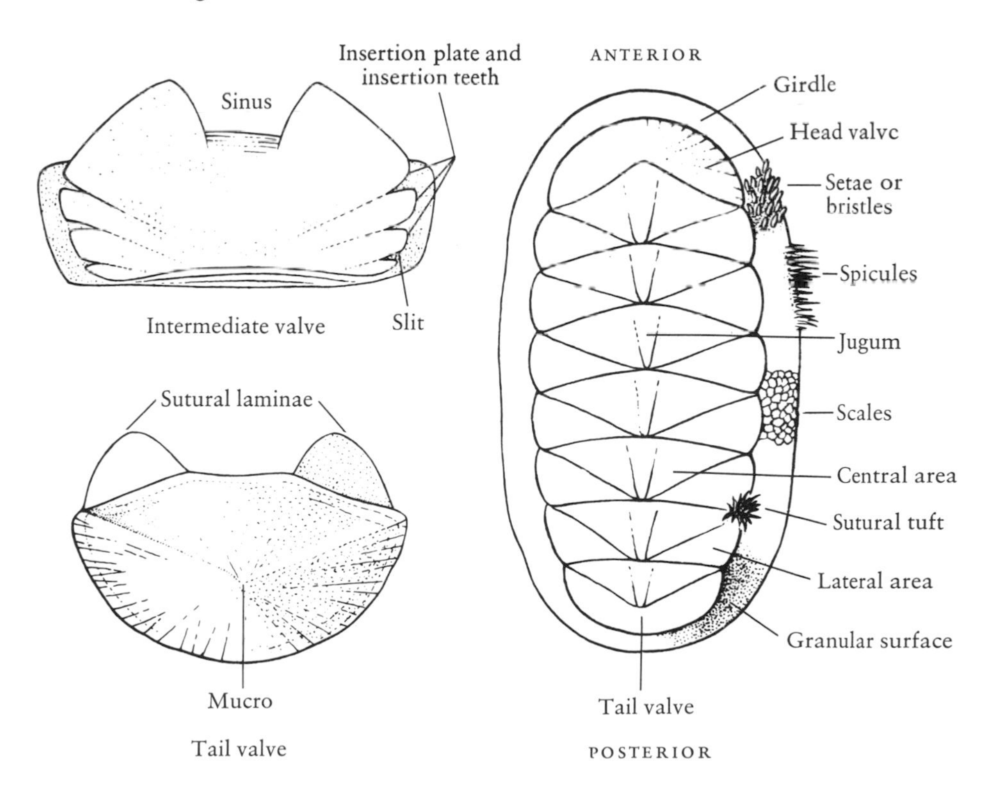

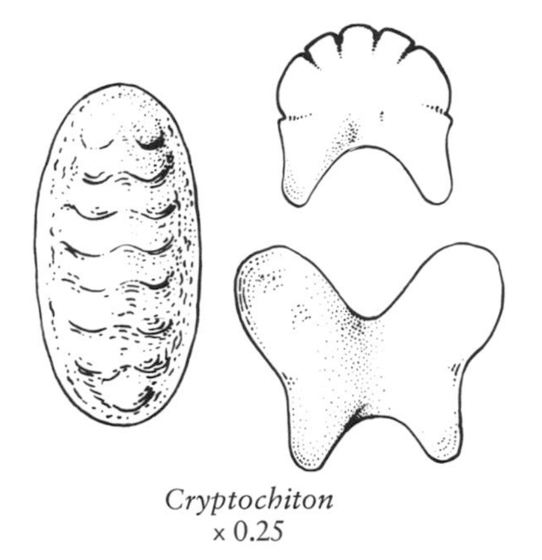

Cryptochiton
x 0.25

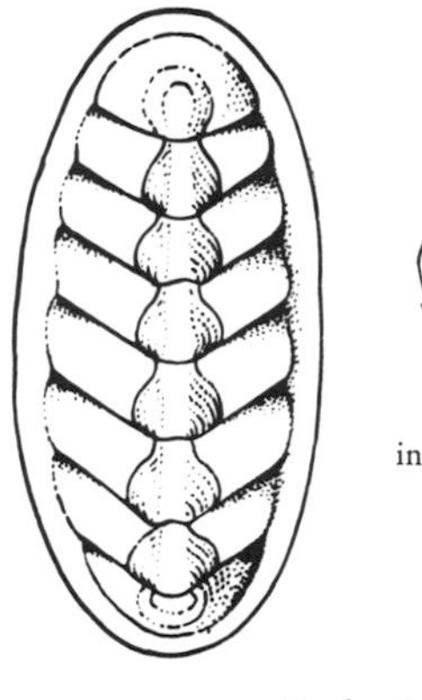

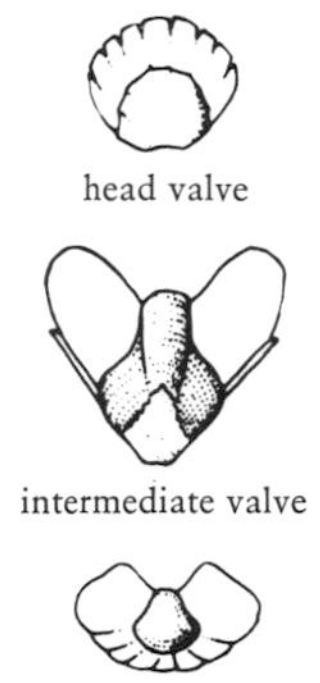

Katharina
x 0.8

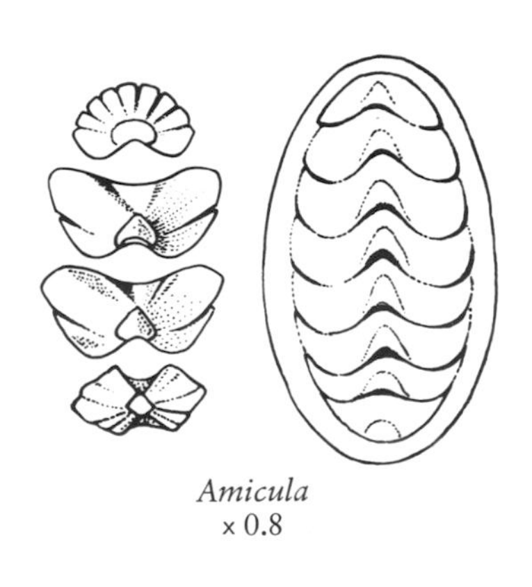

Amicula
x 0.8

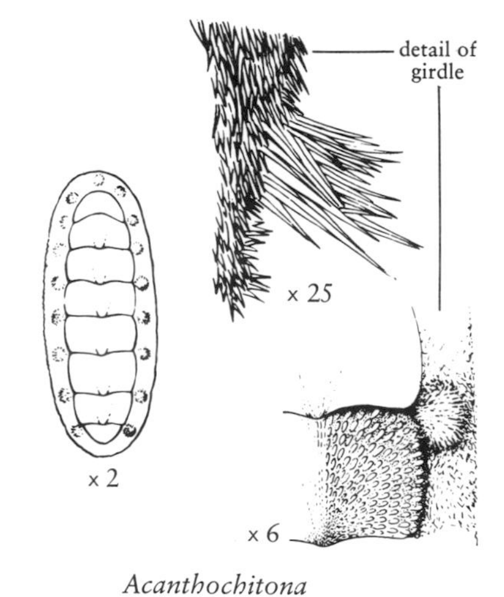

Acanthochitona

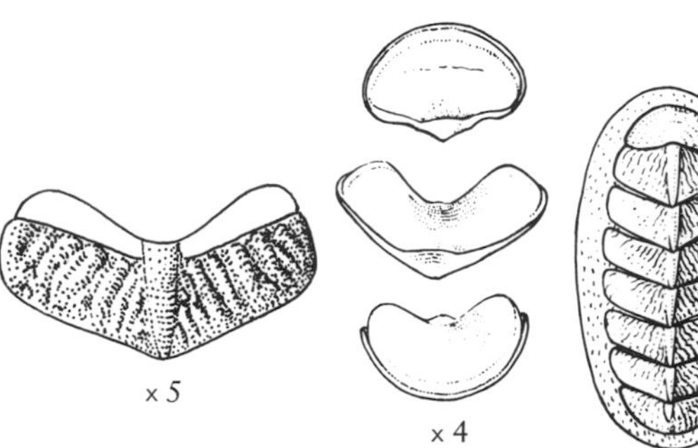

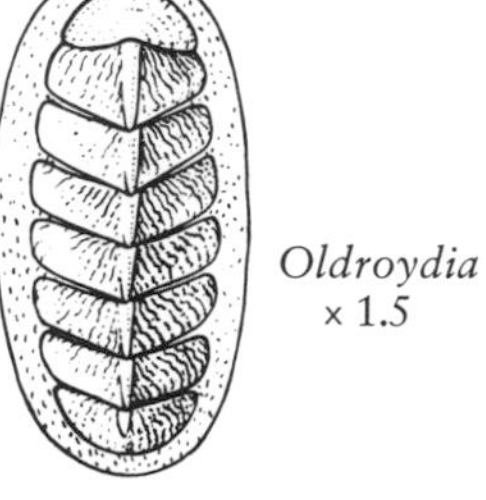

Oldroydia
x 1.5

1 Girdle largely or entirely covering valves 2
Girdle bordering but not widely encroaching on valves 5
2(1) Valves of adult completely covered by girdle . . . . . . *Cryptochiton*
With a small to moderate area of valves always visible 3
3(2) Girdle smooth-surfaced . . . . *Katharina*
Girdle hairy or velvet-surfaced 4
4(3) Exposed shell area small; girdle hair-tufts scattered . . . . . . *Amicula*
Exposed area as wide as girdle; tufts concentrated near sutures . . . . *Acanthochitona*
5(1) Valves separated by narrow intercalations of girdle . . . . *Oldroydia*
Girdle attached to valves only around their ends 6
6(5) Girdle much wider in front, valves short but wide . . . *Placiphorella*
Girdle uniform in width, valves not conspicuously short 7
7(6) Intermediate valves split by a wedge of cartilage at jugum . . . . *Schizoplax*
Jugum entire, not split through the middle 8
8(7) Surface of girdle smooth . . . . *Tonicella*
Surface of girdle roughened by granules ("sandy"), scales, spines, or setae (bristles) 9

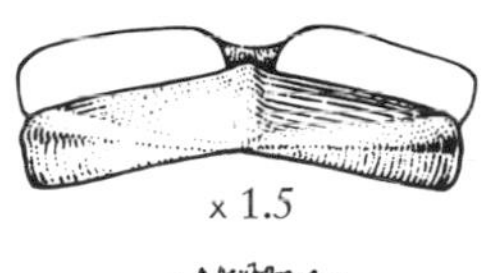

× 1.5

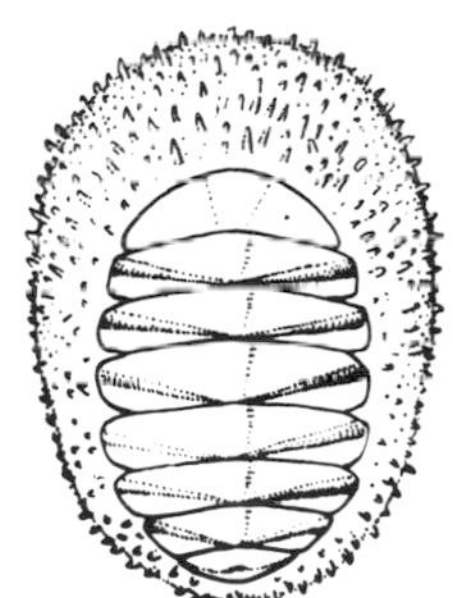

Placiphorella
× 0.7

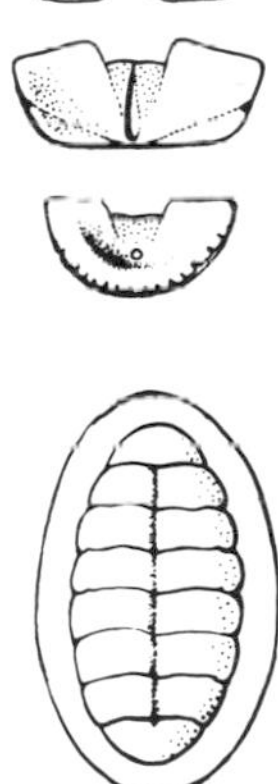

Schizoplax
× 1

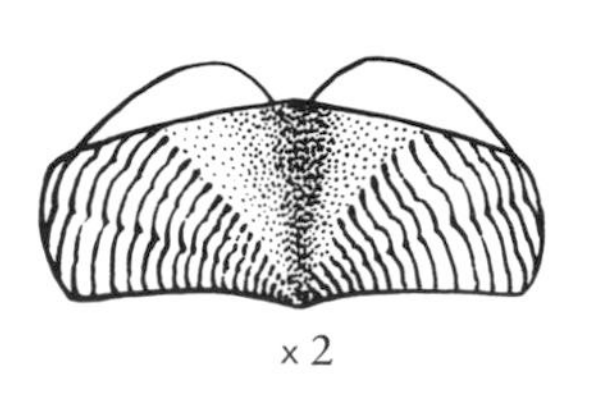

× 2

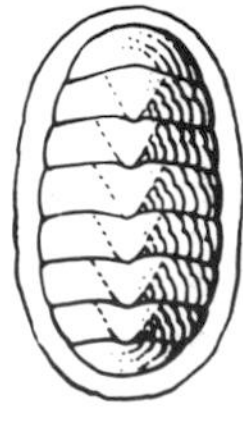

Tonicella
× 1

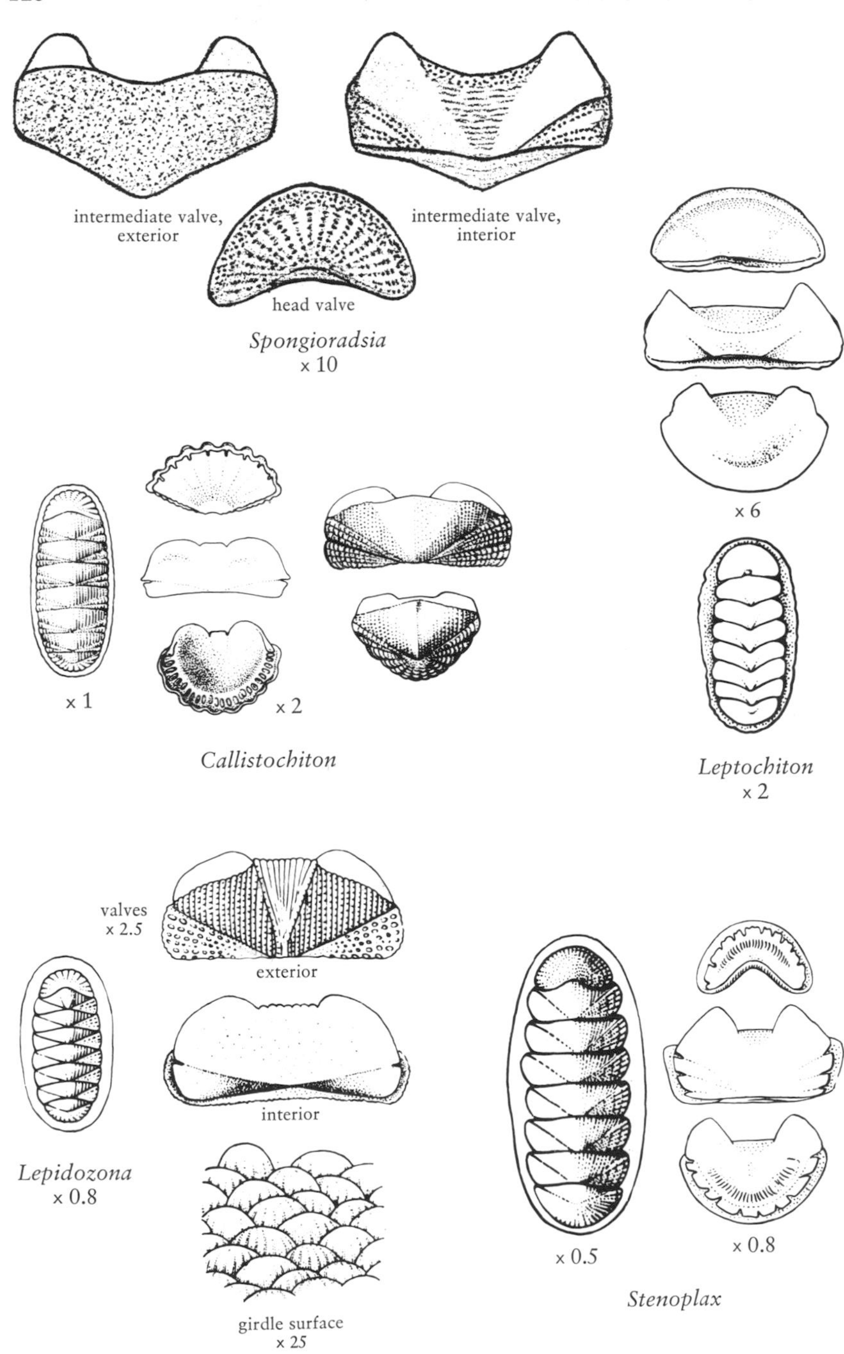

Spongioradsia

Leptochiton

Callistochiton

Lepidozona

Stenoplax

9(8) Valves with articulamentum (i.e. the shell layer that forms the sutural laminae) spongy and porous .... *Spongioradsia*
Valves with articulamentum of normal dense texture 10
10(9) Girdle surface covered with uniform fine chaffy scales 11
Girdle surface not scaly (granular to hairy) 16
11(10) Sinus more than twice as wide as sutural laminae; insertion plates wanting .... *Leptochiton*
Sinus about equal in width to sutural laminae; insertion plates present 12
12(11) With heavy ribs festooning the margins of head and tail valves .... *Callistochiton*
Ribs not affecting the even margins of end valves 13
13(12) Insertion teeth thick at edges .... *Lepidozona*
Insertion teeth sharp-edged 14
14(13) Elongate, relatively large; tail valve larger than head valve... .... *Stenoplax*
Ovate, of medium to small size, head and tail valves equal in size 15
15(14) Tail valve with a raised mucro .... *Ischnochiton*
Tail valve depressed, flat .... *Lepidochitona*

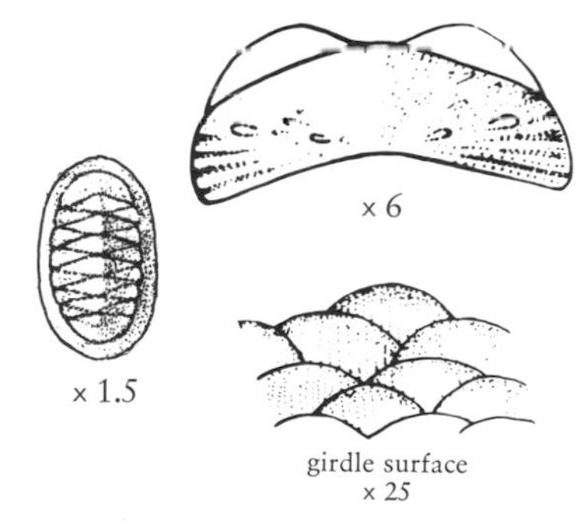

Ischnochiton

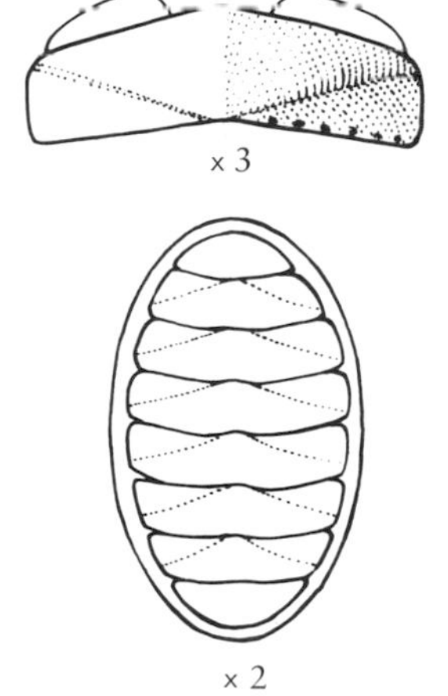

Lepidochitona

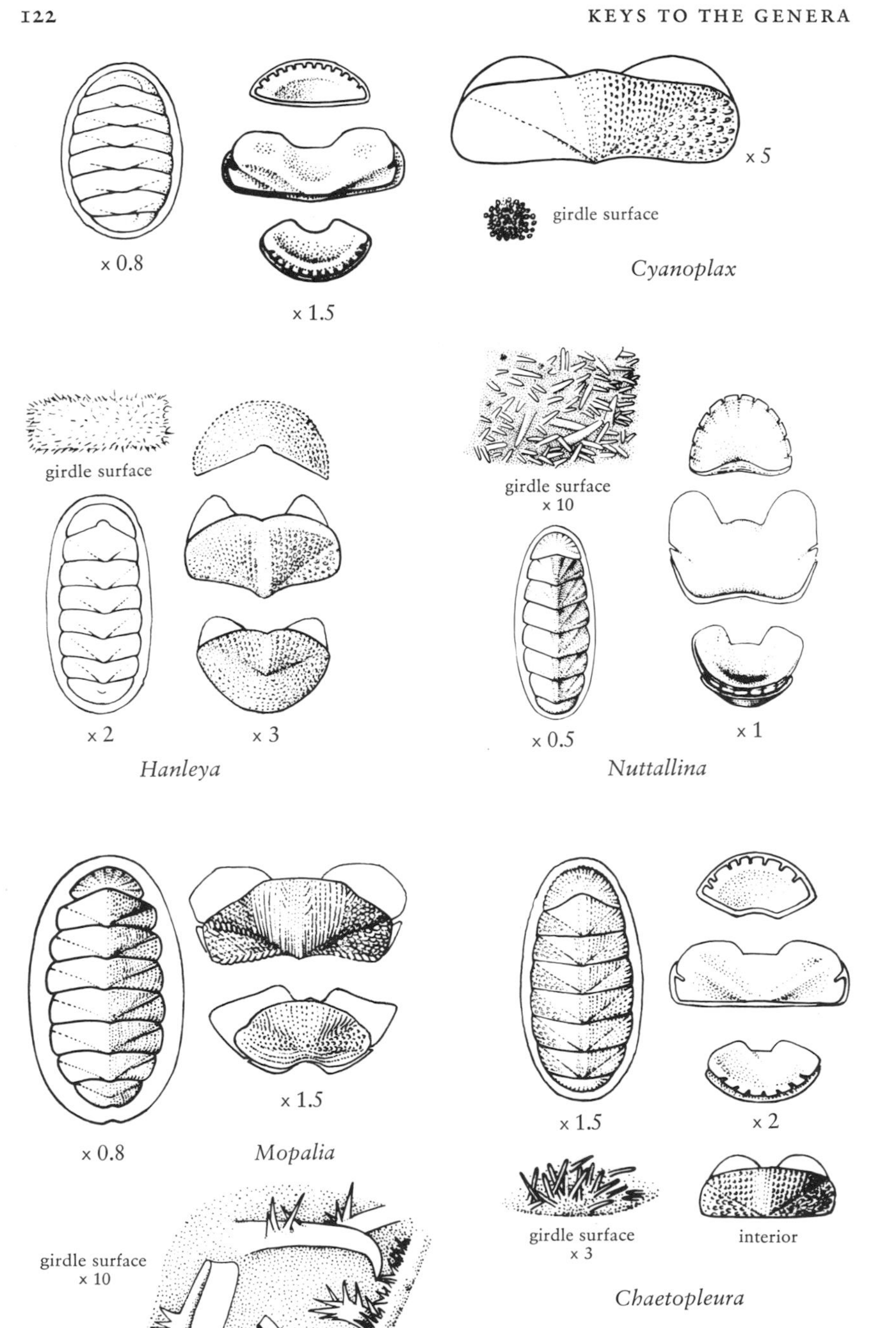
x 0.8
x 1.5
x 5
girdle surface
Cyanoplax
girdle surface
x 2
x 3
Hanleya
girdle surface
x 10
x 0.5
x 1
Nuttallina
x 0.8
x 1.5
Mopalia
girdle surface
x 10
x 1.5
x 2
girdle surface
x 3
interior
Chaetopleura

16(10) Girdle surface finely pebbled; insertion plates crenulate, acute, directed forward . . . . *Cyanoplax*
Girdle surface hairy to spiculose; plates not directed forward . . . 17
17(16) Girdle with spicules or small spines 18
Girdle with hairs or bristles (setae) 19
18(17) Head valve not slit; insertion plates roughened . . . . *Hanleya*
Head valve with nine to twelve slits; insertion plates long, smooth . . . . *Nuttallina*
19(17) Tail valve notched behind, with a single pair of slits at sides . . . . *Mopalia*
Tail valve evenly arched, with more than one pair of slits 20
20(19) Setae simple, scattered; central areas with lines of pustules . . . . *Chaetopleura*
Setae branched; central areas smooth or with irregular granules . . 21
21(20) Central areas of valves smooth . . . . *Basiliochiton*
Central areas sculptured with fine granules . . . . *Dendrochiton*

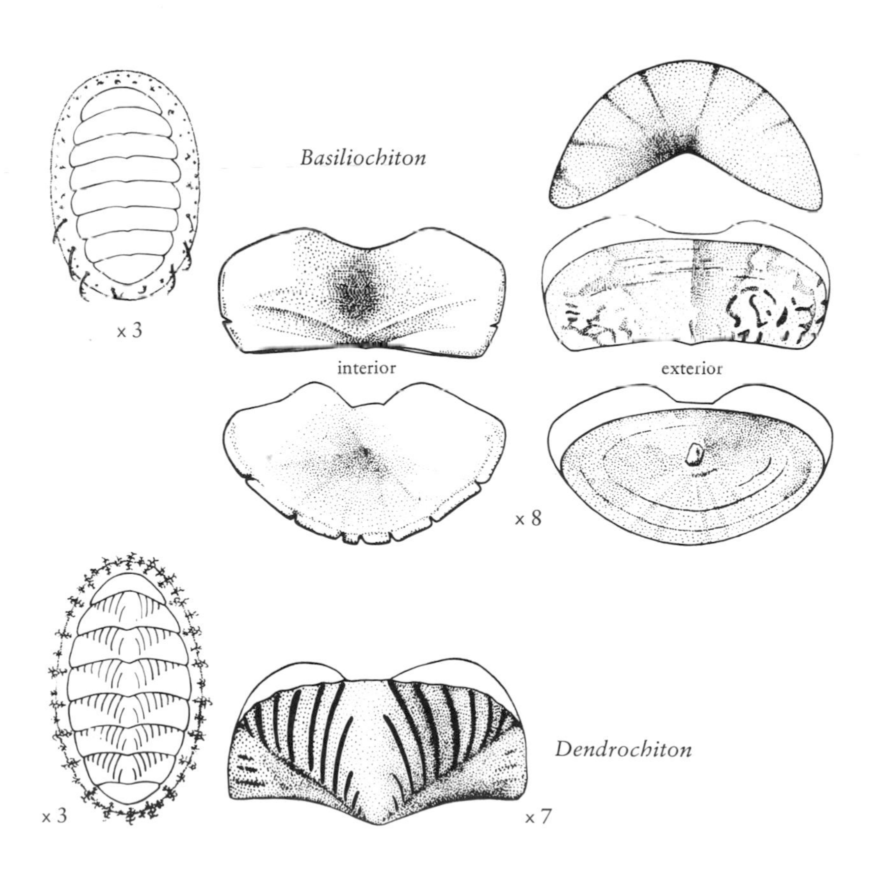

KEY TO THE APLACOPHORA

Because there are no shells in this class—the only stiffening being spicules embedded in the integument or mantle surface—keys to genera are not offered here. However, a key to the families and diagrammatic sketches representing the two orders may be of use. The animals live on the sea floor, mostly in deep water, although a few species have been dredged in water only a few meters deep. Some live on coelenterates; others feed on protozoa or detritus. All West Coast genera are given in the Systematic Lists and under Ranges and Habitats.

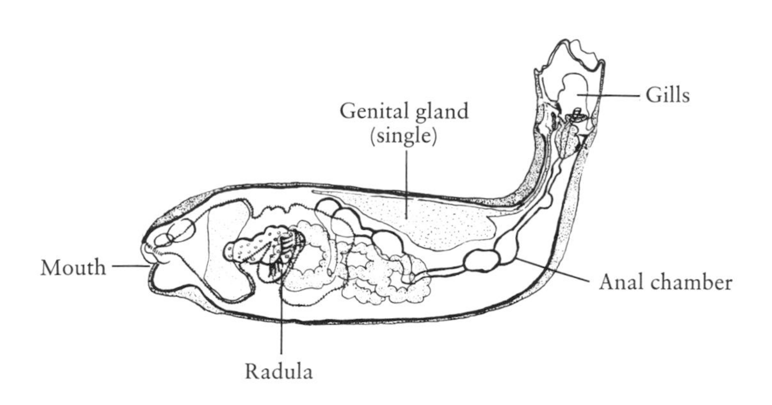

GENERAL ANATOMY
x 25

1 Body spiculose on all surfaces; openings of the mouth and anal chamber terminal; unisexual, with single median gonad . . . . CHAETODERMATIDAE

Body with a ventral groove interrupting the spicules; openings not necessarily terminal; bisexual, with paired genital glands. . . 2

2(1) Epidermis lacking papillae; spicules flat; radula present, its teeth mostly in two rows .... DONDERSIIDAE
Epidermis with papillae; spicules needle-like; radula absent in some 3

3(2) Body crescent-shaped to short-cylindrical; spicules somewhat flattened; radula lacking; gills present .... NEOMENIIDAE
Body elongate; spicules in several layers; radula usually present; gills lacking .... PRONEOMENIIDAE

Family CHAETODERMATIDAE
x 1.5

Family NEOMENIIDAE
x 1.5

KEY TO THE SCAPHOPODA

Scaphopods (tooth or tusk shells) are of simple form, the soft parts being completely enclosed in a tubular shell that is open at both ends. The animals are detritus feeders, burrowing shallowly just below the surface of the sea floor, mostly offshore. Northwest Coast Indians valued the shells as a form of money and devised special brushlike tools to fish for them from boats.

There are numerous species of scaphopods represented on the West Coast, but only four genera, most of which may be divided into subgenera on the basis of shell sculpture. The two families represented are separated on the basis of foot shape and mantle-margin shape.

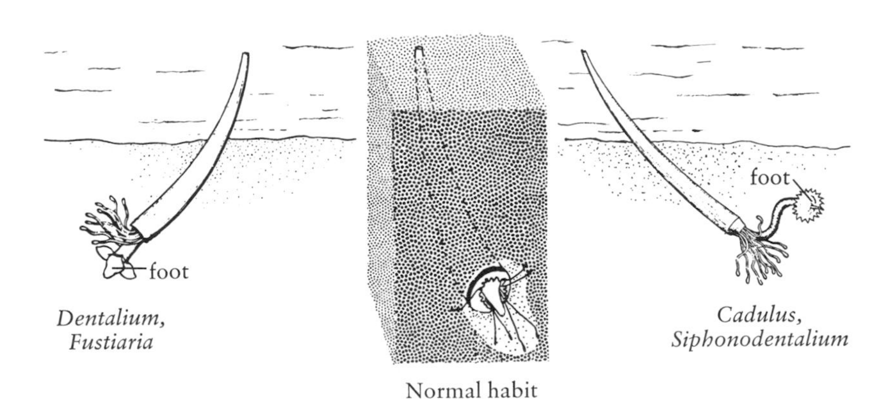

1 Maximum diameter of shell a little posterior to aperture .... *Cadulus*
Maximum diameter of shell at apertural margin 2
2(1) Shell surface longitudinally striate, at least posteriorly .... *Dentalium*
Shell surface smooth throughout 3
3(2) Apex lobed .... *Siphonodentalium*
Apex evenly truncate or with a simple slit .... *Fustiaria*

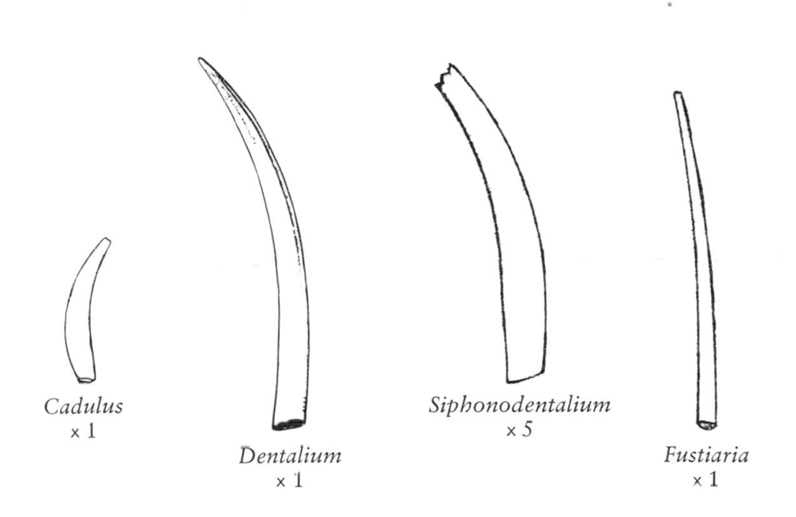

Cadulus × 1

Dentalium × 1

Siphonodentalium × 5

Fustiaria × 1

KEY TO THE CEPHALOPODA

Although the cephalopods of the West Coast are soft-bodied animals, they contain some hard parts that—if isolated from the animal—may be very perplexing. Since no West American form has a true external shell, the word "shell" is used here in quotation marks for want of a better term. The calcareous "shell" of *Argonauta* is a container for eggs, and the corneous structures of other groups are either mandibles that supplement the radula in manipulation of the food material or stiffening rods ("pens") within the body mass. Because these occasional calcareous or corneous structures are generally of questionable value in making determinations, and because the systematics of the class are currently under intensive investigation by a number of specialists, no key will be attempted here. All West Coast genera, however, are given in the Systematic Lists and under Ranges and Habitats.

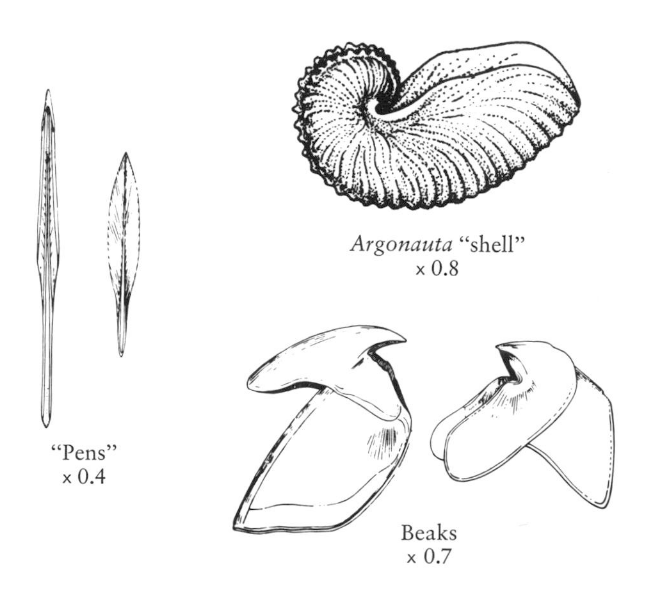

Argonauta "shell"
x 0.8

"Pens"
x 0.4

Beaks
x 0.7

REFERENCE MATTER

SYSTEMATIC LISTS

As stated in the Introduction, these lists serve to show how the keyed genera fit into place in the framework of molluscan classification. A dozen or more classificatory ranks or levels are generally recognized in this formal structure, below the level of the kingdom. The six most commonly used ranks, in descending order, are: phylum, class, order, family, genus, and species. Intermediate ranks may be supplied, if needed, by adding the prefixes "super-" and "sub-"; "super-" seems to be used only at the family level, but "sub-" is used below the class, order, family, genus, and species levels, thus adding another half-dozen categories of rank. The use of italic type sets off the generic- and subgeneric-level names (the subgenera further distinguished by parentheses), but all the rankings higher than that (set in roman type) are specified by name throughout the lists—for example, "Family CARDIIDAE."

Alternative ways of grouping mollusks have been proposed by one author or another, depending on what morphological features they saw as dominant. Thus, especially at the higher levels, there may be competing terms. The best-known of the synonyms are cited here. Usually they have meaning as Latin or Greek words; translations, given in brackets, provide clues to the rationale behind the various proposals. Also in square brackets are notes in explanation of some names (e.g., "[ICZN, pending]"). Explanations of these abbreviations are given in the Introduction (in part), and in the Glossary.

No conventions seem to determine the sequence in which textbooks or manuals treat the several classes. Logically, this might be by appearance in geologic time, but few if any major treatises have taken this approach consistently. Using such a criterion we would find the Gastropoda competing with the Monoplacophora for first place, both having appeared during that dawn period when the soft-bodied animals first developed hard parts that could fossilize—Lower Cambrian time. Pelecypods are next on the stage, also in Lower Cambrian. Cephalopoda and Polyplacophora tie for the next place, appearing in the Upper Cambrian. Scaphopoda were later, though still in the early part of the Paleozoic era, probably in the

Ordovician, positively in the Silurian. No one knows when the Aplacophora came into existence, for they developed no hard parts capable of fossilization.

A different class sequence was established in West Coast literature by Dall in his 1921 checklist: Pelecypoda first, Scaphopoda second, Gastropoda third, and Amphineura (chitons and solenogastres, i.e., the polyplacophorans and aplacophorans) fourth. Many workers have continued using this pattern. However, they have less often followed Dall's strange reversal of internal sequence—simple-to-complex generic sequence for the pelecypods, complex-to-simple for the gastropods. In the present work, the sequence is eclectic, in part following the geologic record, in part being influenced by length of text and grouping of illustrations. Aplacophora are cited next after Polyplacophora because of their previous association in the literature.

The listing of genera within the families and of subgenera within the genera is alphabetical. With minor exceptions, so also is that of subfamilies within families, the exceptions being those groups in which alphabetizing would mask relationships well demonstrated in published systematic reviews. Above the family level, alphabetization ends. The nominate or type family or superfamily customarily is cited first under the name of the next higher category, since it is the radical on which the higher group is based.

For those who are statistically minded, the following quantitative summary of units for the West North American molluscan fauna may be of interest:

Class	Orders	Families	Genera
Gastropoda	15	125	343
Pelecypoda	8	59	173
Polyplacophora	1	8	22
Aplacophora	2	4	11
Scaphopoda	--	2	4
Cephalopoda	4	21	37
TOTAL	30	219	590

CLASS GASTROPODA

["stomach-footed"]

Two systems of classification have been favored by specialists on the Gastropoda. In the one, there is a twofold primary division (into the subclasses Streptoneura and Euthyneura)—as in Pilsbry in Zittel, *Textbook of Paleontology*, Eastman edition (1900), the scheme followed by Taylor and Sohl (1962). In the other, there is a threefold primary division (subclasses Prosobranchia, Opisthobranchia, and

Pulmonata)—as in Thiele, *Handbuch der Systematischen Weichtierkunde* (1935), which is followed by Moore et al. (1960). Although the resulting classifications may employ no actual terms in common, the disparity is not really fundamental, for it reflects more a matter of names to be applied than of groupings to be recognized. The classification adopted here is in the main that of the *Treatise on Invertebrate Paleontology* (Moore et al., 1960), with the alternative terms indicated as synonyms.

Among the Opisthobranchia, the classification of the nudibranchs presents some unusual problems. So many organ systems seem to have classificatory value—buccal apparatus, digestive tract, hepatic glands (liver), ctenidia (gills), and reproductive system—that a complex hierarchy of groupings has developed among the specialists, involving a number of unconventional ranks, such as "tribe," "group," "division," and "subdivision." Taylor and Sohl (1962) attempted to reconcile this hierarchy with the standard categories. The present list accepts their scheme in the main; it retains only one nonconventional term, for a grouping higher than the superfamily but lower than the suborder, the "infraorder." This rank may have some merit in pointing up morphological characters common to two or more superfamilies within a suborder.

Subclass PROSOBRANCHIA
(=PROSOBRANCHIATA) ["gills forward"]
(=STREPTONEURA ["twisted nerves"])

Order ARCHAEOGASTROPODA ["ancient stomach-footed"]
Superfamily PLEUROTOMARIACEA
Family 1. HALIOTIDAE
Haliotis Linnaeus, 1758
Family 2. SCISSURELLIDAE
Scissurella Orbigny, 1824
Sinezona Finlay, 1927
[=*Schismope*, auctt.; *Coronadoa* Bartsch, 1946, juvenile]
Superfamily FISSURELLACEA
Family 3. FISSURELLIDAE
Subfamily FISSURELLINAE
Fissurella Bruguière, 1789
Subfamily EMARGINULINAE
Fissurisepta Seguenza, 1873
Puncturella Lowe, 1827
(Cranopsis) A. Adams, 1860
(Puncturella), s.s.
Rimula Defrance, 1827
Scelidotoma McLean, 1966
[=*Hemitoma*, auctt.]
Subfamily FISSURELLIDINAE
Diodora Gray, 1821
Lucapinella Pilsbry, 1890
Megatebennus Pilsbry, 1890
Megathura Pilsbry, 1890

Superfamily PATELLACEA
Family 4. ACMAEIDAE
Subfamily ACMAEINAE
Acmaea Eschscholtz, 1833
[ICZN Op. 344, 1955]
(Acmaea), s.s.
(Problacmaea) Golikov & Kussakin, 1972
(Rhodopetala) Dall, 1921
(Tectura) Gray, 1847
Subfamily PATELLOIDINAE
Collisella Dall, 1871
Lottia Sowerby, 1834, *ex* Gray MS
Notoacmea Iredale, 1915
Family 5. LEPETIDAE
Cryptobranchia Middendorff, 1851
Lepeta Gray, 1847
(Iothia) Gray, 1850
(Lepeta), s.s.
Superfamily TROCHACEA
Family 6. TROCHIDAE
Subfamily CALLIOSTOMATINAE
Calliostoma Swainson, 1840
(Alertalex) Dell, 1956
(Ampullotrochus) Monterosato, 1890
(Calliostoma), s.s.
(Tristichotrochus) Ikebe, 1942
(Ziziphinus) Gray, 1843

Subfamily Margaritinae
Bathybembix Crosse, 1893
(Bathybembix), s.s.
(Cidarina) Dall, 1909
Calliotropis Seguenza, 1903
Margarites Gray, 1847
(Cantharidoscops) Galkin, 1955
(Margarites), s.s.
(Pupillaria) Dall, 1909
Turcica A. Adams, 1854
Subfamily Solariellinae
Solariella Wood, 1842
(Macheroplax) Friele, 1877
(Minolia) A. Adams, 1860
(Solariella), s.s.
Subfamily Monodontinae
Norrisia Bayle, 1880
Tegula Lesson, 1835
(Agathistoma) Olsson & Harbison, 1953
(Chlorostoma) Swainson, 1840
(Promartynia) Dall, 1909
(Stearnsium) Berry, 1958
Subfamily Halistylinae
Halistylus Dall, 1890
Subfamily Umboniinae
Lirularia Dall, 1909
Family 7. Liotiidae
Liotia Gray, 1847
Macrarene Hertlein & Strong, 1951
[=*Cyclostrema*, auctt., in part, juvenile]
Family 8. Skeneidae
Parviturbo Pilsbry & McGinty, 1945
[=*Arene*, auctt., not H. & A. Adams, 1854]
Family 9. Turbinidae
Subfamily Astraeinae
Astraea Röding, 1798
(Megastraea) McLean, 1970
(Pomaulax) Gray, 1850
Subfamily Homalopomatinae
Homalopoma Carpenter, 1864
(Homalopoma), s.s.
(Panocochlea) Dall, 1908
Moelleria Jeffreys, 1865
Family 10. Phasianellidae
Tricolia Risso, 1826
Superfamily Cocculinacea
Family 11. Cocculinidae
Cocculina Dall, 1882

Order Mesogastropoda ["middle stomach-footed"]
(=Ctenobranchiata ["comb-gills"])
Superfamily Seguenziacea
Family 12. Seguenziidae
Seguenzia Jeffreys, 1876
Superfamily Littorinacea
Family 13. Littorinidae
Littorina Férussac, 1822
(Algamorda) Dall, 1918
(Littorina), s.s.
Family 14. Lacunidae
Aquilonaria Dall, 1886
Haloconcha Dall, 1886
Lacuna Turton, 1827
Superfamily Rissoacea
Family 15. Rissoidae
Subfamily Rissoinae
Alvinia Monterosato, 1884
[=*Alvania*, auctt.]
(Alvinia), s.s.
(Willetia) Gordon, 1939
Subfamily Anabathrinae
Amphithalamus Carpenter, 1864
Anabathron Frauenfeld, 1867
Nannoteretispira Habe, 1961
Subfamily Barleeinae
Barleeia Clark, 1855
(Barleeia), s.s.
(Nodulus) Monterosato, 1878
(Pseudodiala) Ponder, 1967
[=*Diala*, auctt.]
Subfamily Cingulinae
Cingula Fleming, 1818
Elachisina Dall, 1918
Mistostigma Berry, 1947
Family 16. Assimineidae
Assiminea Fleming, 1828
[=*Syncera*, auctt.]
Family 17. Rissoellidae
Rissoella J. E. Gray, 1847
[=*Rissoella* M. E. Gray, 1850, auctt.; *Jeffreysia* Forbes & Hanley, 1850]
Family 18. Rissoinidae
Rissoina Orbigny, 1840
Family 19. Truncatellidae
Cecina A. Adams, 1861[5]
Truncatella Risso, 1826
Family 20. Caecidae
Caecum Fleming, 1813
Elephantanellum Bartsch, 1920
Fartulum Carpenter, 1857

Micranellum Bartsch, 1920
Family 21. SKENEOPSIDAE
Skeneopsis Iredale, 1915
Family 22. VITRINELLIDAE
Cyclostremiscus Pilsbry & Olsson, 1945
[=*Circulus*, auctt.]
Episcynia Mörch, 1875
Leptogyra Bush, 1897
Scissilabra Bartsch, 1907
Solariorbis Conrad, 1865
[=*Delphinoidea*, auctt.]
Teinostoma A. Adams, 1851
(Pseudorotella) Fischer, 1857
(Teinostoma), s.s.
Vitrinella C. B. Adams, 1850
(Docomphala) Bartsch, 1907
(Vitrinella), s.s.
Vitrinorbis Pilsbry & Olsson, 1952
Family 23. CHORISTIDAE
Choristes Carpenter in Dawson, 1872
[may prove to belong in NATICACEA]
Superfamily TURRITELLACEA
Family 24. TURRITELLIDAE
Tachyrynchus Mörch, 1868
[=*Tachyrhynchus*, unjustified emendation, Mörch, 1875]
Turritella Lamarck, 1799
Turritellopsis G. Sars, 1878
Vermicularia Lamarck, 1799
Family 25. VERMETIDAE
Dendropoma Mörch, 1861
[=*Spiroglyphus*, auctt.]
Petaloconchus Lea, 1843
(Macrophragma) Carpenter, 1857
Serpulorbis Sassi, 1827
[=*Aletes* Carpenter, 1857]
Vermetus Daudin, 1800
(Thylaeodus) Mörch, 1860
[=*Bivonia*, auctt.]
Superfamily CERITHIACEA
Family 26. CERITHIIDAE
Subfamily CERITHIOPSINAE
Bittium Gray, 1847
Cerithiopsis Forbes & Hanley, 1851
Metaxia Monterosato, 1884
Seila A. Adams, 1861
Subfamily DIASTOMATINAE
Alaba H. & A. Adams, 1853
Alabina Dall, 1902
Diastoma Deshayes, 1850
Subfamily LITIOPINAE
Litiopa Rang, 1829
Subfamily TRIPHORINAE
[may prove to warrant ranking as a family]
Triphora Blainville, 1828
[=*Trifora*, auctt.]
Family 27. POTAMIDIDAE
Batillaria Benson, 1842[5]
Cerithidea Swainson, 1840
Superfamily EPITONIACEA
Family 28. EPITONIIDAE
Epitonium Röding, 1798
(Asperiscala) De Boury, 1909
(Boreoscala) Kobelt, 1902
(Depressiscala) De Boury, 1909
(Globiscala) De Boury, 1909
(Nitidiscala) De Boury, 1909
Opalia H. & A. Adams, 1853
(Dentiscala) De Boury, 1886
(Nodiscala) De Boury, 1889
(Opalia), s.s.
Family 29. JANTHINIDAE
Janthina Röding, 1798
Superfamily EULIMACEA
Family 30. EULIMIDAE
Balcis Gray, 1847
Eulima Risso, 1826
Haliella Monterosato, 1878
Niso Risso, 1826
Sabinella Monterosato, 1890
Family 31. ACLIDIDAE
Aclis Lovén, 1846
(Aclis), s.s.
(Graphis) Jeffreys, 1867
(Schwengelia) Bartsch, 1947
Superfamily HIPPONICACEA
Family 32. HIPPONICIDAE
Hipponix Defrance, 1819
Sabia Gray, 1847[5]
Family 33. FOSSARIDAE
Fossarus Philippi, 1841[2,3]
Macromphalina Cossmann, 1888
[=*Megalomphalus*, auctt.]
Superfamily CALYPTRAEACEA
Family 34. CALYPTRAEIDAE
Calyptraea Lamarck, 1799
Crepidula Lamarck, 1799
Crepipatella Lesson, 1831

(*Crepipatella*), *s.s.*
(*Verticumbo*) Berry, 1940
Crucibulum Schumacher, 1817
Family 35. CAPULIDAE
Capulus Montfort, 1810
Family 36. TRICHOTROPIDIDAE
Torellia Jeffreys, 1867, *ex* Lovén MS
Trichotropis Broderip & Sowerby, 1829
(*Boetica*) Dall, 1918
(*Iphinoe*) H. & A. Adams, 1854
(*Iphinopsis*) Dall, 1924
(*Trichotropis*), *s.s.*
Superfamily ATLANTACEA (=HETEROPODA ["variously footed"])
Family 37. ATLANTIDAE
Atlanta Lesueur, 1817
Oxygyrus Benson, 1835
Protatlanta Tesch, 1908
Family 38. CARINARIIDAE
Cardiapoda Orbigny, 1835
Carinaria Lamarck, 1801
Family 39. PTEROTRACHEIDAE
Pterotrachea Forskål, 1775
Superfamily NATICACEA
Family 40. NATICIDAE
Subfamily AMPULLOSPIRINAE
Amauropsis Mörch in Rink, 1857
Subfamily NATICINAE
Natica Scopoli, 1777
(*Cryptonatica*) Dall, 1892
[?=*Tectonatica* Sacco, 1890]
Subfamily POLINICINAE
Bulbus Smith, 1838
Calinaticina Burch & Campbell, 1963
[=*Eunaticina*, auctt.]
Neverita Risso, 1826
(*Glossaulax*) Pilsbry, 1929
(*Neverita*), *s.s.*
Polinices Montfort, 1810
(*Lunatia*) Gray, 1847
[?*Euspira* Agassiz in Sowerby, 1838]
Subfamily SININAE
Sinum Röding, 1798
Superfamily LAMELLARIACEA
Family 41. LAMELLARIIDAE
Lamellaria Montagu, 1815
Onchidiopsis Bergh, 1853
Family 42. Velutinidae
Piliscus Lovén, 1859
[=*Capulacmaea* Sars, 1859]
Velutina Fleming, 1820
Superfamily TRIVIACEA
Family 43. TRIVIIDAE
Subfamily TRIVIINAE
Trivia Broderip, 1837
(*Dolichupis*) Iredale, 1930
(*Pusula*) Jousseaume, 1884
Subfamily ERATOINAE
Erato Risso, 1826
(*Erato*), *s.s.*
(*Hespererato*) Schilder, 1932
Superfamily CYPRAEACEA
Family 44. CYPRAEIDAE
Cypraea Linnaeus, 1758
(*Zonaria*) Jousseaume, 1884
Family 45. OVULIDAE (=AMPHIPERATIDAE)
Pedicularia Swainson, 1840
Simnia Risso, 1826
[=*Neosimnia*, auctt.]
(*Delonovula*) Cate, 1973
(*Simnialena*) Cate, 1973
(*Spiculata*) Cate, 1973
Superfamily CYMATIACEA
Family 46. CYMATIIDAE
Cymatium Röding, 1798
Fusitriton Cossmann, 1903
[=*Argobuccinum*, auctt.]
Family 47. BURSIDAE
Bursa Röding, 1798
Order NEOGASTROPODA ["newer stomach-footed"] (=STENOGLOSSA ["narrow tongue"])
Superfamily MURICACEA
Family 48. MURICIDAE
Subfamily MURICINAE
Muricanthus Swainson, 1840[2]
Pterynotus Swainson, 1833
(*Pterochelus*) Jousseaume, 1880
Subfamily MURICOPSINAE
Maxwellia Baily, 1950
Murexiella Clench & Perez-Farfante, 1945
Subfamily OCENEBRINAE
Ceratostoma Herrmannsen, 1846 [ICZN Op. 704, 1964]
?*Eupleura* H. & A. Adams, 1853
Ocenebra Gray, 1847
Pteropurpura Jousseaume, 1880

(*Pteropurpura*), *s.s.*
(*Shaskyus*) Burch & Campbell, 1964
[=*Jaton*, auctt.]
Roperia Dall, 1898
Trachypollia Woodring, 1928
[=*Morula*, auctt., in part; *Morunella* Emerson & Hertlein, 1964: *teste* Radwin & D'Attilio, 1972]
Urosalpinx Stimpson, 1865[5]
Subfamily TROPHONINAE
Austrotrophon Dall, 1902
Trophonopsis Bucquoy, Dautzenberg & Dollfus, 1882
(*Boreotrophon*) Fischer, 1884
(*Nodulotrophon*) Habe & Ito, 1965
(*Trophonopsis*), *s.s.*
Family 49. CORALLIOPHILIDAE
Latiaxis Swainson, 1840
(*Babelomurex*) Coen, 1922
[=*Coralliophila*, auctt.]
Family 50. RAPANIDAE
Forreria Jousseaume, 1880
Family 51. THAIDIDAE
Acanthina Fischer de Waldheim, 1807
Nucella Röding, 1798
[=*Thais*, auctt., in part]
Superfamily BUCCINACEA
Family 52. BUCCINIDAE
Buccinum Linnaeus, 1758
Searlesia Harmer, 1914
Volutharpa Fischer, 1856
Family 53. NEPTUNEIDAE (=CHRYSODOMIDAE)
Ancistrolepis Dall, 1894
(*Ancistrolepis*), *s.s.*
(*Clinopegma*) Grant & Gale, 1931
Beringius Dall, 1886 [ICZN Op. 469, 1957]
Colus Röding, 1798
(*Anomalosipho*) Dautzenberg & Fischer, 1912
(*Aulacofusus*) Dall, 1918
(*Latisipho*) Dall, 1916
Engina Gray, 1839
Exilioidea Grant & Gale, 1931
Kelletia Fischer, 1884
Liomesus Stimpson, 1865
Macron H. & A. Adams, 1853
Mohnia Friele, 1879
Morrisonella Bartsch, 1945
Neptunea Röding, 1798
[=*Chrysodomus* Swainson, 1840]
Plicifusus Dall, 1902
Pyrulofusus Mörch, 1869
Sulcosinus Dall, 1894
Volutopsius Mörch in Rink, 1857
Family 54. COLUMBELLIDAE
Aesopus Gould, 1860
(*Aesopus*), *s.s.*
(*Ithyaesopus*) Olsson & Harbison, 1953
Amphissa H. & A. Adams, 1853
Mitrella Risso, 1826
Nassarina Dall, 1889
(*Zanassarina*) Pilsbry & Lowe, 1932
Family 55. MELONGENIDAE
Busycon Röding, 1798
(*Busycotypus*) Wenz, 1943[5]
Family 56. NASSARIIDAE
Nassarius Duméril, 1806
(*Arcularia*) Link, 1807
(*Caesia*) H. & A. Adams, 1853
(*Demondia*) Addicott, 1965
(*Ilyanassa*) Stimpson, 1865[5]
Family 57. FASCIOLARIIDAE
Subfamily FUSININAE
Fusinus Rafinesque, 1815
Superfamily VOLUTACEA
Family 58. VOLUTIDAE
Arctomelon Dall, 1915
[=*Boreomelon* Dall, 1918]
Sigaluta Rehder, 1967[2]
Tractolira Dall, 1896[2]
Family 59. VOLUTOMITRIDAE
Volutomitra H. & A. Adams, 1853
Family 60. TURBINELLIDAE (=XANCIDAE [ICZN Op. 489, 1957])
Metzgeria Norman, 1879
Ptychatractus Stimpson, 1865
Surculina Dall, 1908
[=*Phenacoptygma* Dall, 1918]
Family 61. OLIVIDAE
Olivella Swainson, 1831
Family 62. MARGINELLIDAE
Subfamily MARGINELLINAE, *s.l.*
Volvarina Hinds, 1844

Subfamily CYSTISCINAE
Cystiscus Stimpson, 1865
Granula Jousseaume, 1875
[=*Kogomea* Habe, 1951]
Granulina Jousseaume, 1888
[=*Cypraeolina* Cerulli-Irelli, 1911; *Merovia* Dall, 1921]
Superfamily MITRACEA
Family 63. MITRIDAE
Mitra Lamarck, 1798
(Atrimitra) Dall, 1918
Superfamily CANCELLARIACEA
Family 64. CANCELLARIIDAE
Admete Möller, 1842
Cancellaria Lamarck, 1799
(Crawfordina) Dall, 1919
(Progabbia) Dall, 1918
Neadmete Habe, 1961
[=*Massyla*, auctt.]
Superfamily CONACEA
Family 65. CONIDAE
Conus Linnaeus, 1758
Family 66. TEREBRIDAE
Terebra Bruguière, 1789
Family 67. TURRIDAE
Subfamily TURRINAE
Cryptogemma Dall, 1918
Subfamily BORSONIINAE
Borsonella Dall, 1908
(Borsonella), s.s.
(Borsonellopsis) McLean, 1971
Ophiodermella Bartsch, 1944
Suavodrillia Dall, 1918
Subfamily CLATHURELLINAE
Clathurella Carpenter, 1857
[ICZN Op. 666, 1962]
(Clathurella), s.s.
(Crockerella) Hertlein & Strong, 1951
Subfamily CLAVINAE
Bellaspira Conrad, 1868
Elaeocyma Dall, 1918
(Elaeocyma), s.s.
(Globidrillia) Woodring, 1928
(Kylix) Dall, 1919
Subfamily CRASSISPIRINAE
Crassispira Swainson, 1840
(Burchia) Bartsch, 1944
(Crassispirella) Bartsch & Rehder, 1939[2]
Subfamily DAPHNELLINAE
Daphnella Hinds, 1844
[=*Eudaphnella* Bartsch, 1933]
Subfamily MANGELIINAE
Cytharella Monterosato, 1875
Mangelia Risso, 1826
(Clathromangelia) Monterosato, 1884
(Glyptaesopus) Pilsbry & Olsson, 1941
(Kurtzia) Bartsch, 1944
(Kurtziella) Dall, 1918
(Kurtzina) Bartsch, 1944
(Mangelia), s.s.
[=*Agathotoma*, auctt., in part]
(Tenaturris) Woodring, 1928
Oenopota Mörch, 1852
[=*Bela*, *Lora*, auctt.]
(Granotoma) Bartsch, 1941
(Nodotoma) Bartsch, 1941
(Obesitoma) Bartsch, 1941
(Oenopota), s.s.
(Propebela) Iredale, 1918
"Pleurotomella" Verrill, 1873[4]
Subfamily MITROLUMNINAE
Cymakra Gardner, 1937
Mitromorpha Carpenter, 1865
Subfamily PSEUDOMELATOMINAE
Pseudomelatoma Dall, 1918
Subfamily TURRICULINAE
Antiplanes Dall, 1903
(Antiplanes), s.s.
(Rectiplanes) Bartsch, 1944
[=*Spirotropis*, auctt.]
(Rectisulcus) Habe, 1958
[=*Taranis*, auctt.]
Carinoturris Bartsch, 1944
Leucosyrinx Dall, 1889
(Aforia) Dall, 1889
(Irenosyrinx) Dall, 1889
(Leucosyrinx), s.s.
(Steiraxis) Dall, 1896
Megasurcula Casey, 1904
Rhodopetoma Bartsch, 1944

Subclass OPISTHOBRANCHIA
(=OPISTHOBRANCHIATA ["gills behind"])
(=EUTHYNEURA ["straight nerves"], in part)

Order PYRAMIDELLIDA (=ENTOMOTAENIATA ["notched fillet"], in part)
Superfamily PYRAMIDELLACEA
Family 68. PYRAMIDELLIDAE
Iselica Dall, 1918
Odostomia Fleming, 1813
(Amaura) Möller, 1842

(Besla) Dall & Bartsch, 1904
(Chrysallida) Carpenter, 1857
(Evalea) A. Adams, 1860
(Evalina) Dall & Bartsch, 1904
(Heida) Dall & Bartsch, 1904
(Iolaea) A. Adams, 1867
(Ivara) Dall & Bartsch, 1903
(Ividella) Dall & Bartsch, 1909
(Menestho) Möller, 1842
(Miralda) A. Adams, 1864
(Odostomia), *s.s.*
(Salassiella) Dall & Bartsch, 1909
Peristichia Dall, 1889
Pyramidella Lamarck, 1799
(Longchaeus) Mörch, 1875
Turbonilla Risso, 1826
(Bartschella) Iredale, 1916 [=*Dunkeria*, auctt., not Carpenter, 1857]
(Chemnitzia) Orbigny, 1839
(Dunkeria) Carpenter, 1857 [=*Pyrgisculus*, auctt.]
(Mormula) A. Adams, 1864
(Pyrgiscus) Philippi, 1841
(Pyrgolampros) Sacco, 1892
(Strioturbonilla) Sacco, 1892
(Turbonilla), *s.s.*
Family 69. CYCLOSTREMELLIDAE
Cyclostremella Bush, 1897
Order PARASITA ["sitting beside," i.e. "guest"]
Family 70. ENTEROXENIDAE
Comenteroxeros Tikasingh, 1961
Thyonicola Mandahl-Barth, 1941
Family 71. ENTOCONCHIDAE
Entocolax Voight, 1888
Entoconcha Müller, 1852
Order CEPHALASPIDEA ["head shield"] (=TECTIBRANCHIATA ["shell-covered gills"], in part)
Superfamily ACTEONACEA
Family 72. ACTEONIDAE
Acteon Montfort, 1810
Microglyphis Dall, 1902 [may prove to belong in the Family RINGICULIDAE]
Rictaxis Dall, 1871
Superfamily BULLACEA
Family 73. BULLIDAE
Bulla Linnaeus, 1758
Family 74. ATYIDAE
Atys Montfort, 1810
Haminoea Turton & Kingston in Carrington, 1830
Family 75. RETUSIDAE
Retusa Brown, 1827 [=*Coleophysis* Fischer, 1883]
Sulcoretusa Burch, 1945 [=*Sulcularia* Dall, 1921, preoccupied]
Volvulella Newton, 1891 [=*Rhizorus*, auctt., not Montfort, 1810]
Superfamily DIAPHANACEA
Family 76. DIAPHANIDAE
Diaphana Brown, 1827
?*Woodbridgea* Berry, 1953 [?=*Brocktonia*, auctt., not Iredale, 1915]
Superfamily PHILINACEA
Family 77. PHILINIDAE
Philine Ascanius, 1772
Family 78. AGLAJIDAE
Aglaja "Renier, 1807" [ICZN, pending]
Chelidonura A. Adams, 1850 [=*Navanax* Pilsbry, 1895]
Family 79. GASTROPTERIDAE
Gastropteron Kosse, 1813, *ex* Meckel MS
Family 80. SCAPHANDRIDAE
Acteocina Gray, 1847
Cylichna Lovén, 1846
Scaphander Montfort, 1810
Family 81. RUNCINIDAE
Runcina Forbes, 1853
Order THECOSOMATA ["sheathed body"] (=PTEROPODA ["winged foot"], in part)
Suborder EUTHECOSOMATA ["true sheathed body"]
Family 82. CAVOLINIIDAE
Cavolinia Abildgaard, 1791 [ICZN Op. 883, 1969] [=*Cavolina* Abildgaard, 1791),
Clio Linnaeus, 1767
Creseis Rang, 1828
Cuvierina Boas, 1886
Diacria Gray, 1840
Hyalocylis Fol, 1875 [=*Hyalocylix*, auctt.]
Styliola Gray, 1850
Family 83. LIMACINIDAE

Limacina Bosc, 1817
[=*Spiratella* Blainville, 1817]
Suborder PSEUDOTHECOSOMATA ["false sheathed body"]
Family 84. CYMBULIIDAE
Corolla Dall, 1871
Cymbulia Péron & Lesueur, 1810
Family 85. DESMOPTERIDAE
Desmopterus Chun, 1889
Family 86. PERACLIDAE
Peracle Forbes, 1844
[=*Peraclis*, auctt.]
Order ANASPIDEA ["without shield"]
Superfamily APLYSIACEA
Family 87. APLYSIIDAF
Subfamily APLYSIINAE
Aplysia Linnaeus, 1767
[=*Tethys*, auctt., ICZN Op. 200, 1954]
(Aplysia), s.s.
(Neaplysia) Cooper, 1863
Subfamily DOLABRIFERINAE
Phyllaplysia Fischer, 1872
[=*Petalifera*, auctt.]
Order GYMNOSOMATA ["naked body"] (=PTEROPODA ["winged foot"], in part)
Family 88. CLIONIDAE
Clione Pallas, 1774
Clionina Pruvot-Fol, 1924
Family 89. CLIOPSIDAE
Cliopsis Troschel, 1854
Family 90. PNEUMODERMATIDAE
Pneumodermopsis Bronn, 1862
[=*Pneumoderma*, auctt.]
Order NOTASPIDEA ["marked shield"]
Superfamily PLEUROBRANCHACEA
Family 91. PLEUROBRANCHIDAE
Subfamily PLEUROBRANCHINAE
Pleurobranchus Cuvier, 1804
Subfamily BERTHELLINAE
Berthella Blainville, 1825
Berthellina Gardiner, 1936
Subfamily PLEUROBRANCHAEINAE
Pleurobranchaea Leue, 1813
Superfamily UMBRACULACEA
Family 92. UMBRACULIDAE
Tylodina Rafinesque, 1819
Order SACOGLOSSA ["shield tongue"] (=ASCOGLOSSA ["bag tongue"])
Superfamily PLAKOBRANCHACEA (=ELYSIACEA, auctt.)
Family 93. PLAKOBRANCHIDAE (=ELYSIIDAE, auctt.)
Elysia Risso, 1818
Family 94. HERMAEIDAE
Alderia Allmann, 1846
Aplysiopsis Deshayes, 1864
[=*Hermaeina* Trinchese, 1874]
(Aplysiopsis), s.s.
(Phyllobranchopsis) Cockerell & Eliot, 1905
Hermaea Lovén, 1844
Stiliger Ehrenberg, 1831
(Placida) Trinchese, 1876
[=*Laura* Trinchese, 1872, not Lacaze-Duthiers, 1865]
(Stiliger), s.s.
Family 95. OLEIDAE
Olea Kjerschow-Agersborg, 1923
Order ACOCHLIDIOIDEA ["kin of those without shells"]
Family 96. HEDYLOPSIDAE
Hedylopsis Thiele, 1931
Order NUDIBRANCHIA (=NUDIBRANCHIATA) ["naked gills"]
Suborder DORIDOIDA (=HOLOHEPATICA ["undivided liver"])
Infraorder CRYPTOBRANCHIA ["hidden gills"]
Superfamily DORIDACEA
Family 97. DORIDIDAE
Subfamily DORIDINAE
Doris Linnaeus, 1758
Glossodoridiformia O'Donoghue, 1927
Subfamily ALDISINAE
Aldisa Bergh, 1878
Rostanga Bergh, 1879
Thordisa Bergh, 1877
Subfamily ARCHIDORIDINAE
Archidoris Bergh, 1878
Atagema Gray, 1850
Subfamily CONUALEVINAE
Conualevia Collier & Farmer, 1964
Subfamily PLATYDORIDINAE
Platydoris Bergh, 1877
Family 98. CHROMODORIDIDAE
Subfamily CHROMODORIDINAE
Chromodoris Alder & Hancock, 1855
[=? *Glossodoris* Ehrenberg, 1831]
Hypselodoris Stimpson, 1855
Subfamily CADLININAE

Cadlina Bergh, 1878 [ICZN Op. 812, 1967]
Subfamily DISCODORIDINAE
Anisodoris Bergh, 1898
Diaulula Bergh, 1879
Discodoris Bergh, 1877
Superfamily GNATHODORIDACEA
Family 99. BATHYDORIDIDAE
Bathydoris Bergh, 1884
Infraorder PHANEROBRANCHIA ["visible gills"]
Superfamily POLYCERATACEA (=NONSUCTORIA ["no buccal bulb"])
Family 100. POLYCERATIDAE
Subfamily POLYCERATINAE
Palio Gray, 1857
Polycera Cuvier, 1817
Subfamily LAILINAE
Crimora Alder & Hancock, 1862
Issena Iredale & O'Donoghue, 1923
Laila MacFarland, 1905
Subfamily TRIOPHINAE
Triopha Bergh, 1880
Family 101. NOTODORIDIDAE
Aegires Lovén, 1844
Superfamily ONCHIDORIDACEA (=SUCTORIA ["buccal bulb"])
Family 102. ONCHIDORIDIDAE
Acanthodoris Gray, 1850
Adalaria Bergh, 1878
Akiodoris Bergh, 1879
Onchidoris Blainville, 1816
Family 103. GONIODORIDIDAE
Ancula Lovén, 1846
Hopkinsia MacFarland, 1905
Okenia Menke, 1830 [ICZN, pending]
Trapania Pruvot-Fol, 1931
[=*Drepania* Lafont, 1874, preoccupied; *Drepanida* MacFarland, 1931]
Family 104. CORAMBIDAE
Corambe Bergh, 1869
Doridella Verrill, 1870
[=*Corambella* Balch, 1899]
Infraorder POROSTOMATA ["pore mouth"]
Superfamily DENDRODORIDACEA
Family 105. DENDRODORIDIDAE
Dendrodoris Ehrenberg, 1831
Doriopsilla Bergh, 1880
Suborder DENDRONOTOIDA
Superfamily DENDRONOTACEA
Family 106. DENDRONOTIDAE
Dendronotus Alder & Hancock, 1845
Family 107. DOTIDAE
Doto Oken, 1815 [ICZN, *nom. conserv.*, Op. 697, 1964]
Family 108. HANCOCKIIDAE
Hancockia Gosse, 1877
Family 109. PHYLLIROIDAE
Cephalopyge Hanel, 1905
Phylliroe Péron & Lesueur, 1810
[=*Phyllirrhoe*, auctt.]
Family 110. TETHYIDAE
Melibe Rang, 1829
[=*Chioraera* Gould, 1852]
Family 111. TRITONIIDAE
Tochuina Odhner, 1963
Tritonia Cuvier, 1797 [ICZN Op. 668, 1963]
[=*Duvaucelia* Risso, 1926, auctt., in part)
Suborder ARMINOIDA
Superfamily ARMINACEA
Family 112. ARMINIDAE
Armina Rafinesque, 1814
Family 113. ANTIOPELLIDAE
Antiopella Hoyle, 1902
Family 114. DIRONIDAE
Dirona Eliot, 1905, *ex* MacFarland MS
Suborder AEOLIDIIDA (=CLADOHEPATICA ["divided liver"], in part)
Infraorder PLEUROPROCTA ["marginal anus"]
Family 115. BABAKINIDAE (=BABAINIDAE, auctt.)
Babakina Roller, 1973
[="*Babaina*" Roller, 1972, preoccupied]
Family 116. FLABELLINIDAE
Coryphella M. E. Gray, 1850 [ICZN Op. 781, 1966]
Flabellinopsis MacFarland, 1966
Infraorder ACLEIOPROCTA ["un-concealed anus"]
Family 117. CUTHONIDAE
Subfamily CUTHONINAE
Cuthona Alder & Hancock, 1855 [ICZN Op. 773, 1966]
Precuthona Odhner, 1929 [ICZN Op. 783, 1966]

Subfamily TERGIPEDINAE
Catriona Winckworth, 1941 [=*Cratena*, auctt., not Bergh, 1864]
Tenellia Costa, 1851
Trinchesia Von Ihering, 1879 [ICZN Op. 777, 1966]
Family 118. EUBRANCHIDAE
Cumanotus Odhner, 1907
Eubranchus Forbes, 1838 [ICZN Op. 774, 1966] [=*Capellinia* Trinchese, 1874]
Family 119. FIONIDAE
Fiona Forbes & Hanley, 1851
Infraorder CLEIOPROCTA ["concealed anus"]
Superfamily AEOLIDIACEA
Family 120. AEOLIDIIDAE
Aeolidia Cuvier, 1797 [ICZN Op. 779, 1966]
Aeolidiella Bergh, 1867[5]
Spurilla Bergh, 1864
Family 121. FACELINIDAE
Emarcusia Roller, 1972
Facelina Alder & Hancock, 1855
Hermissenda Bergh, 1879
Phidiana Gray, 1850
Order GYMNOPHILA ["loving exposure"] (=SOLEOLIFERA ["sandal-bearing"])
Superfamily ONCHIDIACEA
Family 122. ONCHIDIIDAE
Onchidella J. E. Gray in M. E. Gray, 1850 [=*Arctonchis* Dall, 1905]

Subclass PULMONATA ["lunged"]

Order BASOMMATOPHORA ["carrying eyes basally"]
Superfamily MELAMPACEA (=ELLOBIACEA)
Family 123. MELAMPIDAE (=ELLOBIIDAE)
Melampus Montfort, 1810
Pedipes Férussac, 1821
Phytia Gray, 1821?[5]
Superfamily SIPHONARIACEA
Family 124. SIPHONARIIDAE
Siphonaria Sowerby, 1823
Williamia Monterosato, 1884
Family 125. TRIMUSCULIDAE (=GADINIIDAE)
Trimusculus Schmidt, 1818 [=*Gadinia* Gray, 1824]

CLASS PELECYPODA

["hatchet-footed"] (= Bivalvia ["two-shelled"]; Lamellibranchia, Lamellibranchiata ["sheet-gilled"])

Classification of pelecypods is, historically, an even more complex matter than that of gastropods, for paleontologists could construe a great deal about the soft-part morphology of pelecypods using shells alone, and they have not hesitated to erect their own classifications, laying much stress on the hinge and muscle scars. Neontologists, for their part, have not been idle, and have preferred to utilize the gills (hence the name Lamellibranchia, favored by biologists) or even the digestive tract for determining the major divisions. The classification adopted here is essentially that used in the *Treatise on Invertebrate Paleontology* (Moore et al., 1969), but with some rearrangement to preserve the familiar sequence of the classic arrangement of Dall (in Zittel, [1896]-1900). In the newer system, the names of orders are based on the names of contained taxa (carrying the standard practice for family-group names one step higher), instead of on the morphologic criteria previously utilized.

Names of genera and the names and dates for authors have been checked against the catalogue of Vokes (1967), which is the most complete and accurate reference list for pelecypod (bivalve, lamellibranch) genera yet available.

Subclass CRYPTODONTA
["hidden teeth"] (=PROTOBRANCHIA
["first gills"], in part)

Order SOLEMYOIDA (=LIPODONTA
["lacking teeth"])
Superfamily SOLEMYACEA
Family 1. SOLEMYIDAE
Solemya Lamarck, 1818
(Acharax) Dall, 1908
(Petrasma) Dall, 1908

Subclass PALAEOTAXODONTA
["early row-teeth"]
(=PROTOBRANCHIA ["first gills"], in part)

Order NUCULOIDA
Superfamily NUCULACEA
Family 2. NUCULIDAE
Acila H. & A. Adams, 1858
Nucula Lamarck, 1799
Superfamily NUCULANACEA
Family 3. NUCULANIDAE
Malletia DesMoulins, 1832
Neilonella Dall, 1881
Nuculana Link, 1807
[=*Leda* Schumacher, 1817]
Portlandia Mörch, 1857
Sarepta Adams, 1860[2]
Spinula Dall, 1908[2]
Tindaria Bellardi, 1875
Yoldia Möller, 1842
(Cnesterium) Dall, 1898
(Kalayoldia) Grant & Gale, 1931
(Megayoldia) Verrill & Bush, 1897
(Yoldia), s.s.
Yoldiella Verrill & Bush, 1897

Subclass PTERIOMORPHIA
["wing-shaped"]

Order ARCOIDA (=PRIONODONTA ["serrate
teeth"]; EUTAXODONTA ["true
row teeth"])
Superfamily ARCACEA
Family 4. ARCIDAE
Subfamily ARCINAE
Barbatia Gray, 1842
(Acar) Gray, 1857
(Barbatia), s.s.[1]
Subfamily ANADARINAE
Anadara Gray, 1847[1]
(Larkinia) Reinhart, 1935
Bathyarca Kobelt, 1891
Subfamily NOETIINAE
Arcopsis Von Koenen, 1885
[=*Fossularca* Cossmann, 1887]
Family 5. GLYCYMERIDIDAE
Glycymeris Da Costa, 1778
[=*Pectunculus*, auctt.]
(Axinola) Hertlein & Grant, 1972
Family 6. LIMOPSIDAE
Limopsis Sassi, 1827
(Empleconia) Dall, 1908
(Limopsis), s.s.
?Family 7. NUCINELLIDAE
(?=MANZANELLIDAE)
Huxleyia A. Adams, 1860 (April)
[=*Cyrilla* A. Adams, 1860 (June)]
?Family 8. PHILOBRYIDAE
Philobrya Cooper, 1867
[="Carpenter, 1872," auctt.]

Order MYTILOIDA (=DYSODONTA ["poor
teeth"])
Superfamily MYTILACEA
Family 9. MYTILIDAE
Subfamily MYTILINAE
Brachidontes Swainson, 1840
[=*Hormomya* Mörch, 1853]
Mytilus Linnaeus, 1758
Septifer Récluz, 1848
Subfamily CRENELLINAE
Crenella Brown, 1827
Gregariella Monterosato, 1883
[=*Botulina* Dall, 1889]
Megacrenella Habe & Ito, 1965
[=*Solamen*, auctt., not Iredale, 1924]
Musculus Röding, 1798
[=*Modiolaria* Beck, 1838]
(Arvella) Scarlato, 1960, *ex* Bartsch MS
(Musculus), s.s.
(Vilasina) Scarlato, 1960, *ex* Bartsch MS
Subfamily LITHOPHAGINAE
Adula H. & A. Adams, 1857
[=*Botula*, auctt., in part]
Lithophaga Röding, 1798
Subfamily MODIOLINAE
Amygdalum Megerle von Mühlfeld, 1811
Dacrydium Torell, 1859
Ischadium Jukes-Browne, 1905
(Geukensia) Van de Poel, 1959[5]

Modiolus Lamarck, 1799 [*nom. conserv.*, ICZN Op. 325, 1955]
[=*Volsella* Scopoli, 1777]
Superfamily PINNACEA
Family 10. PINNIDAE
Atrina Gray, 1847
Order PTERIOIDA (=PTEROCONCHIDA ["winged shells"])
Superfamily PTERIACEA
Family 11. PTERIIDAE
Isognomon [Lightfoot], 1786[3]
Pteria Scopoli, 1777[3]
[=*Avicula* Bruguière, 1792]
Superfamily OSTREACEA
Family 12. OSTREIDAE
Crassostrea Sacco, 1897[5]
Ostrea Linnaeus, 1758
Superfamily PECTINACEA
Family 13. PECTINIDAE
Argopecten Monterosato, 1889
[=*Plagioctenium* Dall, 1898; *Aequipecten*, auctt.]
Chlamys Röding, 1798
Cyclopecten Verrill, 1897
Delectopecten Stewart, 1930
(Arctinula) Thiele, 1935
(Delectopecten), s.s.
Hinnites Defrance, 1821
Leptopecten Verrill, 1897
Lyropecten Conrad, 1862[1]
Pecten Müller, 1776
(Flabellipecten) Sacco, 1897
(Patinopecten) Dall, 1898
Propeamussium Gregorio, 1884
(Parvamussium) Sacco, 1897
(Polynemamussium) Habe, 1951
Family 14. DIMYIDAE
Dimya Rouault, 1850
Superfamily LIMACEA
Family 15. LIMIDAE
Lima Bruguière, 1797
(Acesta) H. & A. Adams, 1858
(Limaria) Link, 1807
(Limatula) Wood, 1839
(Plicacesta) Vokes, 1963
Superfamily ANOMIACEA
Family 16. ANOMIIDAE
Anomia Linnaeus, 1758
Pododesmus Philippi, 1837
(Monia) Gray, 1850
(Tedinia) Gray, 1853

Subclass HETERODONTA ["differentiated teeth"]
Order VENEROIDA (=TELEODONTA ["perfected teeth"])
Superfamily CRASSATELLACEA
Family 17. CRASSATELLIDAE
Subfamily CRASSATELLINAE
Eucrassatella Iredale 1924
[=*Crassatella, Crassatellites*, auctt.]
Subfamily SCAMBULINAE
Crassinella Guppy, 1874
Family 18. ASTARTIDAE
Astarte J. Sowerby, 1816
(Astarte), s.s.
(Rictocyma) Dall, 1872
(Tridonta) Schumacher, 1817
Superfamily CARDITACEA
Family 19. CARDITIDAE
Subfamily CARDITAMERINAE
Cyclocardia Conrad, 1867
Glans Megerle von Mühlfeld, 1811
Miodontiscus Dall, 1903
Subfamily THECALIINAE
Milneria Dall, 1881
Superfamily GLOSSACEA
Family 20. VESICOMYIDAE
Calyptogena Dall, 1891
Vesicomya Dall, 1886
Superfamily ARCTICACEA
Family 21. BERNARDINIDAE
Bernardina Dall, 1910
Halodakra Olsson, 1961
Superfamily LUCINACEA
Family 22. LUCINIDAE
Subfamily LUCININAE
Epilucina Dall, 1901
Lucina Bruguière, 1797
(Here) Gabb, 1866
(Lucinisca) Dall, 1901
Parvilucina Dall, 1901
Subfamily MYRTAEINAE
Lucinoma Dall, 1901
Family 23. THYASIRIDAE
Adontorhina Berry, 1947
[=*Axinulus* Verrill & Bush, 1898, *teste* Ockelmann, *in litt.*]
Axinopsida Keen & Chavan in Chavan, 1951
[=*Axinopsis* Sars, 1878, not Tate, 1868]
Thyasira Lamarck, 1818

Family 24. UNGULINIDAE
(=DIPLODONTIDAE)
Diplodonta Bronn, 1831
[=*Taras*, auctt.]
Felaniella Dall, 1899
(Zemysia) Finlay, 1927
Superfamily GALEOMMATACEA
Family 25. ERYCINIDAE
Lasaea Brown, 1827
Family 26. KELLIIDAE
Bornia Philippi, 1836
Kellia Turton, 1822
"*Pseudopythina*" Monterosato, 1884[4]
Family 27. LEPTONIDAE
"*Lepton*" Turton, 1822[4]
Family 28. MONTACUTIDAE
Aligena Lea, 1846[1]
Montacuta Turton, 1822
Mysella Angas, 1877 (Aug.)
[=*Rochefortia* Vélain, 1877 (post-Nov.)]
Odontogena Cowan, 1964
Orobitella Dall, 1900
Pristes Carpenter, 1864
[=*Serridens* Dall, 1899]
Tomburchus Harry, 1969
Family 29. GALEOMMATIDAE
Cymatioa Berry, 1964[2]
Superfamily CHLAMYDOCONCHACEA
Family 30. CHLAMYDOCONCHIDAE
Chlamydoconcha Dall, 1884
Superfamily CYAMIACEA
Family 31. SPORTELLIDAE
"*Anisodonta*" Deshayes, 1858[3]
Sportella Deshayes, 1858
Family 32. TURTONIIDAE
Turtonia Alder, 1848
Superfamily CHAMACEA (=PACHYODONTIDA ["thick-toothed"])
Family 33. CHAMIDAE
Chama Linnaeus, 1758
Pseudochama Odhner, 1917
Superfamily CARDIACEA (=CYCLODONTA ["circle teeth"])
Family 34. CARDIIDAE
Subfamily TRACHYCARDIINAE
Trachycardium Mörch, 1853
(Dallocardia) Stewart, 1930
Subfamily FRAGINAE
Americardia Stewart, 1930
Subfamily LAEVICARDIINAE
Clinocardium Keen, 1936
Laevicardium Swainson, 1840
Serripes Gould, 1841
Subfamily PROTOCARDIINAE
Nemocardium Meek, 1876
(Keenaea) Habe, 1951
Superfamily VENERACEA
Family 35. VENERIDAE
Subfamily VENERINAE
Ventricolaria Keen, 1954
[=*Ventricola*, *Antigona*, auctt.]
Subfamily MERETRICINAE
Tivela Link, 1807
(Pachydesma) Conrad, 1854
Transennella Dall, 1883
Subfamily PITARINAE
Amiantis Carpenter, 1864
Pitar Römer, 1857
[=*Pitaria*, auctt.]
Saxidomus Conrad, 1837
Subfamily GEMMINAE
Gemma Deshayes, 1853[5]
Subfamily CLEMENTIINAE
Compsomyax Stewart, 1930
[=*Marcia*, auctt., in part]
Subfamily TAPETINAE
Irus Schmidt, 1818
[=*Notirus*, *Venerupis* (in part), auctt.]
(Irusella) Hertlein & Grant, 1972
Liocyma Dall, 1870
Psephidia Dall, 1902
Tapes Megerle von Mühlfeld, 1811
(Ruditapes) Chiamenti, 1900[5]
Subfamily CHIONINAE
Chione Megerle von Mühlfeld, 1811
(Chione), *s.s.*
(Chionista) Keen, 1958
Humilaria Grant & Gale, 1931
[=*Marcia*, auctt., in part]
Mercenaria Schumacher, 1817?[1]
Protothaca Dall, 1902
[=*Tapes*, *Paphia*, *Venerupis*, auctt.]
(Callithaca) Dall, 1902
(Protothaca), *s.s.*
Family 36. PETRICOLIDAE
Petricola Lamarck, 1801
(Petricola), *s.s.*
(Petricolaria) Stoliczka, 1870[2,5]
Family 37. COOPERELLIDAE

Cooperella Carpenter, 1864
Superfamily MACTRACEA
Family 38. MACTRIDAE
Subfamily MACTRINAE
Mactra Linnaeus, 1767
(Mactra), s.s.
(Mactrotoma) Dall, 1894
(Micromactra) Dall, 1894
Spisula Gray, 1837
(Mactromeris) Conrad, 1868
(Symmorphomactra) Dall, 1894
Subfamily LUTRARIINAE
Tresus Gray, 1853 (early Jan.)
[=*Schizothaerus* Gray, 1853 (late Jan.)]
Subfamily PTEROPSELLINAE
Raeta Gray, 1853[1]
Superfamily TELLINACEA
Family 39. TELLINIDAE
Leporimetis Iredale, 1930
(Florimetis) Olsson & Harbison, 1953
[=*Apolymetis*, auctt.]
Macoma Leach in Ross, 1819
(Macoma), s.s.
(Psammacoma) Dall, 1900
(Rexithaerus) Tryon, 1869, *ex* Conrad MS
Tellina Linnaeus, 1758
(Angulus) Megerle von Mühlfeld, 1811
[=*Moerella, Oudardia*, auctt.]
(Cadella) Dall, Bartsch & Rehder, 1938
(Megangulus) Afshar, 1969
(Peronidia) Dall, 1900
(Tellinella) Mörch, 1853
Family 40. DONACIDAE
Donax Linnaeus, 1758
Family 41. PSAMMOBIIDAE (=GARIDAE)
Subfamily PSAMMOBIINAE
Gari Schumacher, 1817
(Gobraeus) Brown, 1844
Heterodonax Mörch, 1853
Subfamily SANGUINOLARIINAE
Nuttallia Dall, 1898
[=*Sanguinolaria*, auctt.]
Family 42. SOLECURTIDAE
Tagelus Gray, 1847
(Mesopleura) Conrad, 1867
(Tagelus), s.s.
Family 43. SEMELIDAE
Abra Lamarck, 1818[2]
Cumingia Sowerby, 1833
Semele Schumacher, 1817
Theora H. & A. Adams, 1856[5]
Superfamily SOLENACEA
Family 44. SOLENIDAE
Subfamily SOLENINAE
Solen Linnaeus, 1758
Subfamily CULTELLINAE
Ensis Schumacher, 1817
Siliqua Megerle von Mühlfeld, 1811
Order MYOIDA (=ASTHENODONTA ["degenerate teeth"])
Suborder MYINA
Superfamily MYACEA
Family 45. MYIDAE
Cryptomya Conrad, 1848
Mya Linnaeus, 1758
(Arenomya) Winckworth, 1930
(Mya), s.s.
Platyodon Conrad, 1837
Sphenia Turton, 1822
Family 46. CORBULIDAE
Corbula Bruguière, 1797
[=*Aloidis* Megerle von Mühlfeld, 1811]
Family 47. SPHENIOPSIDAE
Grippina Dall, 1912
Superfamily GASTROCHAENACEA
Family 48. GASTROCHAENIDAE
Gastrochaena Spengler, 1783[1]
Superfamily HIATELLACEA
Family 49. HIATELLIDAE
Cyrtodaria Reuss, 1801, *ex* Daudin MS
Hiatella Bosc, 1801, *ex* Daudin MS
[=*Saxicava* Fleuriau de Bellevue, 1802]
Panomya Gray, 1857
Panopea Ménard, 1807 [ICZN, pending]
[=*Panope*, auctt.]
Saxicavella Fischer, 1878
Suborder PHOLADINA
Superfamily PHOLADACEA
Family 50. PHOLADIDAE
Subfamily PHOLADINAE
Barnea Risso, 1826
Zirfaea Gray, 1842
Subfamily JOUANNETIINAE
Netastoma Carpenter, 1864
[=*Nettastomella* Carpenter,

1865, unneeded new name]
Subfamily MARTESIINAE
Chaceia Turner, 1955
Parapholas Conrad, 1849 [?1848]
Penitella Valenciennes, 1846 [=*Navea* Gray, 1851; *Pholadidea*, auctt.]
Subfamily (XYLOPHAGAINAE (=XYLOPHAGINAE, auctt., preoccupied *fide* Turner in Moore (1969))
Xylophaga Turton, 1822
Xyloredo Turner, 1972
Family 51. TEREDINIDAE
Subfamily TEREDININAE
Lyrodus Binney, 1870, *ex* Gould MS[5]
Teredo Linnaeus, 1758?[5]
Subfamily BANKIINAE
Bankia Gray, 1842

Subclass ANOMALODESMATA (=ANOMALODESMACEA) ["irregular hinge"]

Order PHOLADOMYOIDA (=EUDESMODONTIDA ["primitive hinge teeth"])
Superfamily PANDORACEA
Family 52. PANDORIDAE
Pandora Bruguière, 1797
Family 53. LATERNULIDAE
Laternula Röding, 1798[5]
Family 54. LYONSIIDAE
Entodesma Philippi, 1845
Lyonsia Turton, 1822
Mytilimeria Conrad, 1837
Family 55. PERIPLOMATIDAE
Periploma Schumacher, 1817
Family 56. THRACIIDAE
Asthenothaerus Carpenter, 1864
Cyathodonta Conrad, 1849
Thracia Sowerby, 1823, *ex* Leach MS
Superfamily POROMYACEA (=SEPTIBRANCHIA ["partitioned gills"])
Family 57. POROMYIDAE
Poromya Forbes, 1844
(Dermatomya) Dall, 1899
(Poromya), s.s. [=*Cetoconcha* Dall, 1886, auctt., in part]
Family 58. CUSPIDARIIDAE
Cardiomya A. Adams, 1864
Cuspidaria Nardo, 1840
Myonera Dall, 1886
Plectodon Carpenter, 1864
Family 59. VERTICORDIIDAE
Halicardia Dall, 1895[2]
Lyonsiella G. Sars, 1872, *ex* M. Sars MS
Policordia Dall, Bartsch & Rehder, 1938 [=*Lyonsiella*, auctt., in part]
Verticordia Sowerby, 1844, *ex* Wood MS

CLASS POLYPLACOPHORA

["bearing many plates"] (= Loricata ["clad in mail"]; Amphineura ["double-nerved"], in part)

The classification adopted here is essentially that of Smith in Moore et al. (1960), with some rearrangement of parts. These are of course the chitons, formerly combined with the aplacophorans as Amphineura by many authors.

Order NEOLORICATA ["new loricates"]
Suborder ACANTHOCHITONINA ["spiny tunic or garment"]
Family 1. ACANTHOCHITONIDAE
Acanthochitona Gray, 1821
Cryptochiton Middendorff, 1847 [=*Amicula*, auctt., in part]
Suborder ISCHNOCHITONINA ["thin tunic"]
Family 2. ISCHNOCHITONIDAE
Basiliochiton Berry, 1918 [?=*Trachydermon* auctt.; ?*Mopaliella* Thiele, 1909, in part]
Cyanoplax Pilsbry, 1892
Ischnochiton Gray, 1847
(Radsiella) Pilsbry, 1892 [ranked as separate genus by some authors]

Lepidochitona Gray, 1821
Lepidozona Pilsbry, 1892
Spongioradsia Pilsbry, 1894
Stenoplax Dall, 1879, *ex* Carpenter MS
Tonicella Carpenter, 1873
Family 3. CALLISTOPLACIDAE
Callistochiton Dall, 1897, *ex* Carpenter MS
Nuttallina Dall, 1871
Family 4. CHAETOPLEURIDAE
Chaetopleura Shuttleworth, 1853
[=*Pallochiton* Dall, 1879]
Family 5. MOPALIIDAE
Amicula Gray, 1847
[=*Symmetrogephyrus* Middendorff, 1847]
Dendrochiton Berry, 1911
Katharina Gray, 1847
Mopalia Gray, 1847
Placiphorella Dall, 1879, *ex* Carpenter MS
Family 6. SCHIZOPLACIDAE
Schizoplax Dall, 1878
Suborder LEPIDOPLEURINA ["scaly sides"]
Family 7. LEPIDOPLEURIDAE
Leptochiton Gray, 1847
Oldroydia Dall, 1894
Family 8. HANLEYIDAE
Hanleya Gray, 1857

CLASS APLACOPHORA

["bearing no plates"] (=Amphineura ["double-nerved"], in part)

This group has been combined with the chitons as Amphineura by many authors, but the relationship is not now thought to be close. The aplacophorans, or solenogastres, as they are more commonly called, are wormlike mollusks that lack shells. They have a spiculose outer layer or integument and (in common with the chitons and the monoplacophoran genus *Neopilina*) a ladderlike nervous system. However, in other ways the solenogastres are highly specialized and unlike other mollusks. They are in the main inhabitants of deep water, where they apparently work over the bottom sediment or feed on coelenterates.

The two orders that are represented on the West Coast are differentiated by the presence or absence of a longitudinal furrow on the ventral surface; the furrow is present in the Neomeniida, interrupting the spiculose integument, whereas in the Chaetodermatida the integument is spiculose throughout.

The principal review of the West American solenogastres is by Heath (1911); Schwabl (1963) has contributed further useful work.

Order NEOMENIIDA ["kin of the new moon," i.e., "crescent-shaped"]
Family 1. NEOMENIIDAE
Alexandromenia Heath, 1911
Pachymenia Heath, 1911
Platymenia Schwabl, 1961
Family 2. DONDERSIIDAE
Dondersia Hubrecht, 1888
Heathia Thiele, 1913
[=*Ichthyomenia*, auctt.]
Nematomenia Simroth, 1893
[=*Herpomenia* Heath, 1911]
Family 3. PRONEOMENIIDAE
Dorymenia Heath, 1911
Proneomenia Hubrecht, 1880
Order CHAETODERMATIDA ["hairy-skinned"]
Family 4. CHAETODERMATIDAE
Chaetoderma Lovén, 1845
[ICZN, Op. 764, 1966]
[=*Crystallophrisson* Möbius, 1875]
Limifossor Heath, 1904
Prochaetoderma Thiele, 1902

CLASS SCAPHOPODA

["shovel-footed"]

This small class has been revised by Ludbrook [in Moore et al. (1960)] and by Emerson (1962). The list of genera for Western North America is a short one, and the class itself is so small that no ranks above family are needed.

Family 1. DENTALIIDAE
Dentalium Linnaeus, 1758
Fustiaria Stoliczka, 1868

Family 2. SIPHONODENTALIIDAE
Cadulus Philippi, 1844
Siphonodentalium Sars, 1859

CLASS CEPHALOPODA

["head-footed"]

Most modern forms of cephalopods lack shells, and the few that have them (for example, *Nautilus* and *Spirula*) do not occur in the Eastern Pacific. Two important earlier papers on the subject are those by Clarke (1966) and Berry (1912).

The outline used here essentially follows that of Young (1972), with some rearrangement and additions.

Subclass COLEOIDEA
["sheathed"]

Order SEPIOIDEA ["cuttlefish kin"]
Family 1. SEPIOLIDAE
Rossia Owen in Ross, 1835
Order TEUTHOIDEA ["squid kin"]
Suborder MYOPSIDA ["nearsighted"]
Family 2. LOLIGINIDAE
Loligo Schneider, 1784
[="Lamarck, 1798," auctt.]
Suborder OEGOPSIDA ["open-sighted"]
Family 3. BATHYTEUTHIDAE
Bathyteuthis Hoyle, 1885
Family 4. CHIROTEUTHIDAE
Chiroteuthis Orbigny, 1839 [?1841]
Valbyteuthis Joubin, 1931
Family 5. CRANCHIIDAE
Subfamily CRANCHIINAE
Cranchia Leach, 1817
Crystalloteuthis Chun, 1906
Leachia Lesueur, 1821
Subfamily TAONIINAE
Galiteuthis Joubin, 1898
Helicocranchia Massy, 1907
Taonius Steenstrup, 1861
Family 6. ENOPLOTEUTHIDAE
Subfamily ENOPLOTEUTHINAE
Abraliopsis Joubin, 1896
Subfamily PYROTEUTHINAE
Pterygioteuthis H. Fischer, 1896
Pyroteuthis Hoyle, 1904
Family 7. GONATIDAE
Berryteuthis Naef, 1921
Gonatopsis Sasaki, 1920
(Boreoteuthis) Nesis, 1971
Gonatus Gray, 1849
Family 8. GRIMALDITEUTHIDAE
Grimalditeuthis Joubin, 1898
Family 9. HISTIOTEUTHIDAE
Histioteuthis Orbigny, 1840
[=*Calliteuthis* Verrill, 1880; *Meleagroteuthis* Pfeffer, 1900]
Family 10. MASTIGOTEUTHIDAE
Mastigoteuthis Verrill, 1881
Family 11. OCTOPOTEUTHIDAE
Octopoteuthis Ruppell, 1844
Family 12. OMMASTREPHIDAE
Dosidicus Steenstrup, 1857
Ommastrephes Orbigny, 1835
[=*Ommatostrephes*, auctt.]
Symplectoteuthis Pfeffer, 1900
Todarodes Steenstrup, 1880
Family 13. ONYCHOTEUTHIDAE
Moroteuthis Verrill, 1881
Onychoteuthis Lichtenstein, 1818

Order OCTOPODA ["eight-footed"]
  Suborder CIRRATA (=CIRROMORPHA ["curly"])
    Family 14. OPISTHOTEUTHIDAE
      Opisthoteuthis Verrill, 1883
    Family 15. STAUROTEUTHIDAE
      Grimpoteuthis Robson, 1932
  Suborder INCIRRATA ["not curly"]
    Family 16. ALLOPOSIDAE
      Alloposus Verrill, 1880
    Family 17. ARGONAUTIDAE
      Argonauta Linnaeus, 1758
    Family 18. BOLITAENIDAE
      Eledonella Verrill, 1884
      Japetella Hoyle, 1885
    Family 19. OCTOPODIDAE
      Benthoctopus Grimpe, 1921
      Octopus Cuvier [1797] [*nom. conserv.*, ICZN Op. 233, 1954]
        [=*Octopodia, Polypus*, auctt.]
    Family 20. OCYTHOIDAE
      Ocythoe Rafinesque, 1814
Order VAMPYROMORPHA ["batlike"]
    Family 21. VAMPYROTEUTHIDAE
      Vampyroteuthis Chun, 1903
        [=*Cirroteuthis*, auctt.]

RANGES AND HABITATS

This tabulation furnishes basic ecological data on all genera treated in this work, to the extent we have been able to uncover this material from the literature. All groups, whether shelled or unshelled, are included. Arrangement is alphabetical by class. Synonyms mentioned in the keys or the Systematic Lists are not included here, nor are subgenera. The superscripted numerals occasionally following generic names apply to certain distributional categories explained in the Introduction. The subscripts "o" and "p," respectively, identify opisthobranchs and pulmonates; the unsubscripted genera are prosobranchs. The exact number of species to be recognized in each genus may be open to debate, but a statement of at least an approximate number should be useful; this is given in the second column. Genera occurring only north of Puget Sound, Washington, are so indicated by the letter "N" in the "Range" column, and those occurring only south of Point Conception, California, by an "S." A dash indicates that the range is central or that it is wide enough to overlap northward or southward. "Int." in the last column indicates an intertidal habitat. Bathymetric records are given only with respect to occurrences of these groups on the Western North American coast; in other provinces the upper or lower depth limits may differ substantially.

Generic unit	*No. of species (approx.)*	*Range*	*Depth and habitat*
		GASTROPODA	
Acanthina	4	--	Int., on rocks
Acanthodoris$_o$	9	--	Int. to 215 m, in rocky areas, on mud in bays, or on docks
Aclis	5	S	Int., on rocks
Acmaea	4	--	Int. to 64 m, on rocks, shells, or coralline algae
Acteocina$_o$	7	--	Int. to 295 m, on sand
Acteon$_o$	4	--	Int. to 75 m, on sand or mud
Adalaria$_o$	3	N	10 to 25 m
Admete	6	--	20 to 730 m
Aegires$_o$	1	--	Int. to 30 m, in rocky areas or on docks
Aeolidia$_o$	1	--	Int. to 870 m, in rocky areas or on mud in bays
Aeolidiella$_o$[5]	1	--	Int. to 5 m, in rocky areas or on docks
Aesopus	4	--	Int. to 40 m
Aglaja$_o$	3-4	--	Int. to 100 m, on mud, sand, or docks
Akiodoris$_o$	1	N	Int.

Generic unit	No. of species (approx.)	Range	Depth and habitat
Alaba	1-2	S	Int. to 45 m, among rocks
Alabina	3	--	Int. to 20 m
Alderia$_o$	1	--	Int., on the alga *Vaucheria* or in *Salicornia* marshes
Aldisa$_o$	2	--	Int. to 10 m, in rocky areas
Alvinia	15	--	Int. to 230 m
Amauropsis	1	N	Int. to 1,300 m, on mud
Amphissa	6	--	Int. to 640 m
Amphithalamus	2	--	Int. to 130 m, among rocks
Anabathron	1	N	Not determined
Ancistrolepis	3	--	65 to 2,500 m
Ancula$_o$	2	--	Int. to 5 m, in rocky areas or on pilings
Anisodoris$_o$	1	--	Int. to 255 m, in rocky areas or on pilings or docks
Antiopella$_o$	2	--	Int. to 295 m, in rocky areas, on mud in bays, or on docks
Antiplanes	8	--	50 to 1,830 m
Aplysia$_o$	3-4	--	Int. to 30 m, in rocky areas or on mudflats
Aplysiopsis$_o$	2	--	Int. to 10 m, in muddy bays or rocky areas, or on kelp holdfasts
Aquilonaria	1	N	Int. (?) to 145 m
Archidoris$_o$	3	--	Int. to 255 m, in rocky areas or on docks
Arctomelon	1	N	60 to 250 m
Armina$_o$	1	--	Int. to 230 m, on muddy sand
Assiminea	1	--	High-tide line, salt marshes
Astraea	3	--	Int. to 75 m, on rocks
Atagema$_o$	1	--	Int. to 210 m, in rocky areas
Atlanta	6-7	--	Pelagic
Atys$_o$	1	S	About 110 m
Austrotrophon	3	S	55 to 135 m
Babakina$_o$	1	S	Int. to 5 m, in rocky areas
Balcis	27	--	Int. to 230 m
Barleeia	6	--	Int. to 50 m
Bathybembix	2	--	50 to 2,930 m
Bathydoris$_o$	1	--	"Deep water"
Batillaria[5]	2	--	Int., on mudflats; introduced from Japan
Bellaspira	1	S	25 to 45 m
Beringius	7	N	35 to 600 m
Berthella$_o$	2	--	Int. to 45 m, in rocky areas
Berthellina$_o$	1	S	Int. to 10 m, in rocky areas
Bittium	13	--	Int. to 660 m
Borsonella	5	--	75 to 1,465 m
Buccinum	26	--	10 to 2,930 m
Bulbus	1	N	35 to 315 m, on mud or sand
Bulla$_o$	2	S	Int. to 45 m, on mudflats
Bursa	1	--	20 to 110 m
Busycon[5]	1	--	Int. to 20 m, in mud of bays; introduced from Atlantic
Cadlina$_o$	6	--	Int. to 220 m, in rocky areas or on docks
Caecum	2	--	Int. to 100 m, among rubble
Calinaticina	1	--	20 to 365 m, on sandy mud

Generic unit	*No. of species (approx.)*	*Range*	*Depth and habitat*
Calliostoma	12	--	Int. to 915 m
Calliotropis	3	--	1,500 to 2,900 m
Calyptraea	3	--	Int. to 140 m, on rocks
Cancellaria	4	--	25 to 550 m
Capulus	1	S	35 to 45 m, on the clam *Pecten diegensis*
Cardiapoda	1	--	Pelagic
Carinaria	1	--	Pelagic
Carinoturris	1	--	550 to 1,100 m
Catriona$_\circ$	3	--	Int. to 20 m, on gravel or on docks
Cavolinia$_\circ$	1	--	Offshore, pelagic
Cecina[5]	1	N	Int., in *Salicornia* marshes; introduced from Northwest Pacific
Cephalopyge$_\circ$	1	--	Pelagic
Ceratostoma	2	--	Int. to 65 m, on rocks
Cerithidea	1	--	Int. on mudflats
Cerithiopsis	16	--	Int. to 200 m
Chelidonura$_\circ$	2	--	Int. to 35 m, on mud, sand, or rocks, especially in bays
Choristes	1	--	2,700 to 3,400 m
Chromodoris$_\circ$	1-2	--	Int. to 10 m, in rocky areas
Cingula	7	--	Int. to 230 m
Clathurella	9	--	Int. to 145 m
Clio$_\circ$	3	--	Offshore, pelagic
Clione$_\circ$	2	--	Offshore, pelagic
Clionina$_\circ$	?	--	Offshore, pelagic
Cliopsis$_\circ$	?	--	Offshore, pelagic
Cocculina	2	--	185 to 1,000 m
Collisella	10	--	Int. to 10 m, on rocks or algae
Colus	20	--	18 to 2,010 m
Comenteroxeros$_\circ$	1	--	Int. to 20 m, parasitic
Conualevia$_\circ$	1	--	Int. to 20 m, in rocky areas
Conus	1	--	Int. to 45 m
Corambe$_\circ$	1	--	Int. to 5 m, on eelgrass, kelp, or the bryozoan *Membranipora*
Corolla$_\circ$	2	--	Offshore, pelagic
Coryphella$_\circ$	9	--	Int. to 100 m, in rocky areas, on offshore kelp, or on docks
Crassispira	2	--	25 to 35 m
Crepidula	12	--	Int. to 165 m, on rocks or shells
Crepipatella	3	--	Int. to 185 m, on rocks or shells
Creseis$_\circ$	1	--	Offshore, pelagic
Crimora$_\circ$	1	--	Int. in rocky areas
Crucibulum	1	--	Int. to 25 m, on rocks or pilings
Cryptobranchia	1	N	Low int. to 60 m
Cryptogemma	1	--	550 to 4,020 m
Cumanotus$_\circ$	1	--	Int. to 5 m, on hydroids or on mudflats or docks
Cuthona$_\circ$	1	--	Int.
Cuvierina$_\circ$	1	--	Offshore, pelagic
Cyclostremella$_\circ$	2	--	20 to 55 m
Cyclostremiscus	1	S	Int. (?) to 10 m
Cylichna$_\circ$	6	--	5 to 200 m

Generic unit	No. of species (approx.)	Range	Depth and habitat
Cymakra	3	--	Int. to 40 m
Cymatium	1	S	About 75 m
Cymbulia$_{o}$	1	S	Offshore, pelagic
Cypraea	1	--	Int. to 45 m, on rocks
Cystiscus	2	--	Int. to 60 m, in rocky areas
Cytharella	1	N	20 to 25 m
Daphnella	2	--	10 to 185 m
Dendrodoris$_{o}$	2	--	Int. to 10 m, in rocky areas
Dendronotus$_{o}$	7	--	Int. to 400 m, in rocky areas, on mud or kelp offshore, on eelgrass, or on docks
Dendropoma	1	--	Int. to 20 m, attached, in dense colonies or corroding a channel in other shells, especially the abalone *Haliotis*
Desmopterus$_{o}$	1	S	Offshore, pelagic
Diacria$_{o}$	1	S	Offshore, pelagic
Diaphana$_{o}$	3	--	Int. to 230 m, in rocky areas
Diastoma	3	S	Int.
Diaulula$_{o}$	1	--	Int. to 35 m, in rocky areas or on docks
Diodora	2	--	Int. to 65 m
Dirona$_{o}$	3	--	Int. to 55 m, in rocky areas, on gravel or mud, or on docks
Discodoris$_{o}$	1	--	Int. to 10 m, in rocky areas
Doridella$_{o}$	1	--	Int. to 5 m, on docks or kelp, or on the bryozoan *Membranipora*
Doriopsilla$_{o}$	1	--	Int. to 45 m, in rocky areas
Doris$_{o}$	2-3	--	Int., in rocky areas
Doto$_{o}$	3-5	--	Int. to 60 m, in rocky areas or on docks, or on hydroids
Elachisina	2-3	S	25 to 35 m
Elaeocyma	4	S	Int. to 90 m
Elephantanellum	1	S	Int.
Elysia$_{o}$	1	--	Int. to ?5 m, in rocky areas, on mudflats, or on algae
Emarcusia$_{o}$	1	--	Int. to 5 m, on hydroids
Engina	1	S	45 to 90 m
Entocolax$_{o}$	1	N	Parasitic in holothurian intestines
Entoconcha$_{o}$	1	N	Parasitic circumorally in holothurians
Episcynia	1	S	4 to 5 m
Epitonium	18	--	Int. to 365 m, often associated with sea anemones
Erato	2	--	Int. to 90 m
Eubranchus$_{o}$	3	--	Int. to 10 m, on rocks or algae, on docks, or offshore on kelp
Eulima	4	--	35 to 110 m
Eupleura	1	S	25 m
Exilioidea	2	--	55 to 625 m
Facelina$_{o}$	2	S	Int., on sand in bays
Fartulum	2	--	Int. to 60 m
Fiona$_{o}$	1	--	Pelagic, on floating wood
Fissurella	1	--	Int. to 40 m, on rocks
Fissurisepta	1	N	860 to 880 m
Flabellinopsis$_{o}$	1	--	Int. to 40 m, in rocky areas

Generic unit	*No. of species (approx.)*	*Range*	*Depth and habitat*
Forreria	1	S	Int. to 135 m
Fossarus[2, 3]	1	S	Int.
Fusinus	6	--	Int. to 185 m
Fusitriton	1	--	Int. to 135 m
Gastropteron$_o$	1-2	--	Offshore, pelagic or in depths of 90 to 245 m
Glossodoridiformia$_o$	1	S	Int., in rocky areas
Granula	2	--	Int. to 100 m
Granulina	1	--	Int. to 110 m
Haliella	2	S	510 to 1,465 m
Haliotis	7	--	Int. to 90 m
Halistylus	1	--	20 to 75 m
Haloconcha	1	N	Int.
Haminoea$_o$	3	--	Int. to 20 m, on rocks or algae
Hancockia$_o$	1	--	Int. to 5 m, in rocky areas or on kelp
Hedylopsis$_o$	1	N	Int.
Hermaea$_o$	1	--	Int. to 5 m, in rocky areas or on the alga *Codium*
Hermissenda$_o$	1	--	Int. to 35 m, on rocks, algae, mudflats, or docks
Hipponix	2	--	Int. to 75 m, cemented to rocks
Homalopoma	8	--	Int. to 90 m, in rocky areas
Hopkinsia$_o$	1	--	Int. to 5 m, in rocky areas
Hyalocylis$_o$	1	S	Offshore, pelagic
Hypselodoris$_o$	2	--	Int. to 30 m, in rocky areas
Iselica$_o$	3	--	Int. to 55 m
Issena$_o$	1	N	Int.
Janthina	1	--	Pelagic
Kelletia	1	S	Int. to 35 m
Lacuna	6	--	Int. to 365 m
Laila$_o$	1	--	Int. to 35 m, in rocky areas
Lamellaria	4	--	Int. to 25 m, associated with ascidians
Latiaxis	1	S	Int. to 45 m
Lepeta	3	--	18 to 200 m
Leptogyra	1	N	Not determined
Leucosyrinx	6	--	1,465 to 2,010 m
Limacina$_o$	1	--	Offshore, pelagic
Liomesus	3	N	20 to 220 m
Liotia	1	--	Int. to 45 m, in rocky areas
Lirularia	10	--	Int. to 90 m, in rocky areas
Litiopa	1	--	Pelagic
Littorina	7	--	Int., near high-tide line
Lottia	1	--	Int., on exposed rocks
Lucapinella	1	--	Int., on rocks
Macrarene	2	--	20 to 35 m
Macromphalina	1	S	20 to 35 m
Macron	1	S	Int., in rocky areas
Mangelia	12	--	Int. to 1,500 m
Margarites	18	--	Int. to 230 m
Maxwellia	1	S	Int. to 55 m, in rocky areas
Megasurcula	2	--	20 to 365 m
Megatebennus	1	--	Int. to 20 m, on rocks
Megathura	1	--	Int. to 10 m, on rocks
Melampus$_p$	1	--	Above high-tide line in bays

Generic unit	*No. of species (approx.)*	*Range*	*Depth and habitat*
Melibe$_o$	1	--	Int. to 30 m, on eelgrass or kelp
Metaxia	1	--	Int. to 55 m, in rocky areas
Metzgeria	1	--	25 to 90 m
Micranellum	1	--	Int. to 75 m
Microglyphis$_o$	2	--	120 to 1,830 m
Mistostigma	1-2	S	90 m (?)
Mitra	2-3	--	Int. to 185 m, in rocky areas
Mitrella	6	--	Int. to 1,200 m
Mitromorpha	1	--	Int. to 45 m, in rocky areas
Moelleria	2	N	Low int. to 230 m
Mohnia	6	--	1,000 to 3,290 m
Morrisonella	1	--	2,930 m
Murexiella	1	S	35 to 70 m, in rocky areas
Nannoteretispira	1	S	Int. to 10 m, in rocky areas
Nassarina	1	S	Int. to 45 m, in rocky areas
Nassarius	10	--	Int. to 365 m, on sand or mud
Natica	1	--	Int. to 1,645 m, on mud
Neadmete	2	--	18 to 1,205 m
Neptunea	12	--	Int. to 1,555 m
Neverita	4	--	Int. to 2,815 m, on mud or sand
Niso	1-2	S	90 m
Norrisia	1	--	Int., on rocks or algae
Notoacmea	6	--	Int., on rocks or algae
Nucella	4	--	Int. to 15 m, in rocky areas
Ocenebra	14	--	Int. to 150 m, in rocky areas
Odostomia$_o$	152	--	Int. to 640 m
Oenopota	62	--	5 to 1,555 m
Okenia$_o$	3	--	Int. to 5 m, on rocks near mud or on docks
Olea$_o$	1	N	Int. to 10 m, on mud in bays or on cephalaspidean egg masses
Olivella	3	--	Int. to 90 m, in sand
Onchidella$_o$	1	--	Int., near high-tide line on rocks
Onchidiopsis	1	N	Int. to 165 m
Onchidoris$_o$	4	--	Int. to 30 m, in rocky areas, on mudflats, or on docks
Opalia	5	--	Int. to 100 m, in rocky areas; may be associated with anemones
Ophiodermella	4	--	9 to 200 m
Oxygyrus	1	S	Pelagic
Palio$_o$	1	N	About 20 m
Parviturbo	2	--	Int. to 45 m, in rocky areas
Pedicularia	1	--	20 to 90 m, on the hydrocoral *Allopora*
Pedipes$_p$	2	S	Int., in rubble near high-tide line
Peracle$_o$	2	S	Offshore, pelagic
Peristichia$_o$	1	S	Int. to 45 m
Petaloconchus	1	--	Int. to 45 m, attached to rocks or other shells
Phidiana$_o$	1	--	Int. to 215 m, in rocky areas
Philine$_o$	4-5	--	45 to 250 m
Phyllaplysia$_o$	1	--	Int. to 5 m, on eelgrass in bays
Phylliroe$_o$	1	S	Offshore, pelagic
Phytia$_p$[5]?	1	--	Int., near high-tide line of bays

Generic unit	*No. of species (approx.)*	*Range*	*Depth and habitat*
Piliscus	1	N	Int. (?) to 165 m
Platydoris$_o$	1	--	30 to 155 m
Pleurobranchaea$_o$	1	--	5(?) to 365 m, on sand or mud
Pleurobranchus$_o$	1	--	Int. to 10 m, in rocky areas
"*Pleurotomella*"[4]	1	S	1,095 m
Plicifusus	10	--	20 to 1,095 m
Pneumodermopsis$_o$	1	--	Offshore, pelagic
Polinices	7	--	Int. to 2,820 m, on mud or sand
Polycera$_o$	4	--	Int. to 60 m, in rocky areas, on mud, or on docks
Precuthona$_o$	1	--	Int. to 20 m, in rocky areas or on mud
Protatlanta	1	S	Pelagic
Pseudomelatoma	3	--	Int. to 75 m
Pteropurpura	4	--	Int. to 120 m
Pterotrachea	1	S	Pelagic
Pterynotus	1	S	200 m
Ptychatractus	2	--	75 to 1,465 m
Puncturella	9	--	Int. to 200 m, on rocks
Pyramidella$_o$	1	S	Int. to 35 m, on eelgrass
Pyrulofusus	4	N	5 to 2,000 m
Retusa$_o$	3	N	20 to 550 m
Rhodopetoma	2	S	145 to 1,095 m
Rictaxis$_o$	1	--	Int. to 85 m, on mudflats or in sand
Rimula	1	S	15 to 45 m
Rissoella	1	S	Not determined
Rissoina	7	--	Int. to 200 m
Roperia	1	S	Int. to 10 m
Rostanga$_o$	1	--	Int. to 20 m, in rocky areas
Runcina$_o$	1	--	Int., on the alga *Endocladia* or the annelid *Phragmatopoma*
Sabinella	1	--	1,500 to 2,930 m
Scaphander$_o$	1	N	Not determined
Scelidotoma	1	--	10 to 100 m, on rocks
Scissilabra	1	--	15 to 25 m
Scissurella	4	--	15 to 1,280 m
Searlesia	1	--	Int., on rocks
Seguenzia	3	--	220 to 1,465 m
Seila	1	--	Int. to 65 m, in rocky areas
Serpulorbis	1	--	Int., cemented to rocks, shells, or pilings
Sigaluta[2]	1	S	3,775 to 3,795 m
Simnia	5	--	20 to 90 m, on gorgonians
Sinezona	1	--	Int. to 35 m, in rocky areas
Sinum	1	--	Int. to 170 m, on sand or mud
Siphonaria$_p$	2	--	Int. to 45 m
Skeneopsis	1	N	Not determined
Solariella	5	--	15 to 1,095 m
Solariorbis	1	S	Not determined
Spurilla$_o$	2	--	Int. to 20 m, in rocky areas or on docks
Stiliger$_o$	2	--	Int., on algae in bays, on docks, or in rocky areas
Styliola$_o$	1	--	Offshore, pelagic
Suavodrillia	2-3	N	20 m (?)
Sulcoretusa$_o$	1	--	About 20 m

Generic unit	*No. of species (approx.)*	*Range*	*Depth and habitat*
Sulcosinus	1	N	640 m
Surculina	1	S	1,190 m
Tachyrynchus	5	--	Int. to 275 m
Tegula	8	--	Int. to 25 m, on rocks
Teinostoma	1	--	Int. to 65 m
Tenellia$_o$	1	--	Int., on docks or pilings
Terebra	2	S	Int. to 55 m
Thordisa$_o$	1	--	Int. to 35 m, in rocky areas
Thyonicola$_o$	1	--	1 to 20 m, parasitic on holothurians
Tochuina$_o$	1	--	Int. to 360 m, in rocky areas
Torellia	2	N	75 m
Trachypollia	1	S	45 m
Tractolira[2]	1	S	3,775 to 3,795 m
Trapania$_o$	1	--	Int., in rocky areas or on docks or pilings
Trichotropis	9	--	1 to 1,190 m
Tricolia	4	--	Int. to 45 m, on algae
Trimusculus$_p$	1	--	Int., on rocks
Trinchesia$_o$	7	--	Int. to 30 m, in rocky areas
Triopha$_o$	6-8	--	Int. to 80 m, in rocky areas, on docks, on mudflats, or on offshore kelp
Triphora	7	--	Int. to 100 m, in rocky areas
Tritonia$_o$	3-5	--	Int. to 1,260 m, in rocky areas or on sand
Trivia	3	--	Int. to 90 m, in rocky areas
Trophonopsis	21	--	Int. to 1,060 m
Truncatella	1	S	Near high-tide line in rocky areas
Turbonilla$_o$	115	--	Int. to 135 m
Turcica	1	--	20 to 65 m
Turritella	2	--	20 to 185 m
Turritellopsis	1	--	20 to 35 m
Tylodina$_o$	1	S	Int. to 25 m, on sponges in rocky areas
Urosalpinx[5]	1	--	Int., on rocks in bays; introduced from Atlantic
Velutina	8	--	Int. to 230 m
Vermetus	1	--	Int., cemented to rocks or shells
Vermicularia	1	S	Int. to 10 m, attached to rocks, in rubble
Vitrinella	7	--	Int. to 50 m
Vitrinorbis	1	S	Int.
Volutharpa	2	N	300 to 600 m
Volutomitra	1	--	150 to 1,465 m
Volutopsius	6	N	Int. to 1,465 m
Volvarina	1	--	Int. to 20 m
Volvulella$_o$	3	--	2 to 550 m
Williamia$_p$	2	--	Int. to 135 m
?Woodbridgea$_o$	1	S	90 m

PELECYPODA

Generic unit	*No. of species (approx.)*	*Range*	*Depth and habitat*
Abra[2]	1	S	3,500 m; not yet reported from California
Acila	1	--	10 to 1,280 m
Adontorhina	1	S	10 to 640 m
Adula	3	--	Int. to 25 m, boring into rocks or nestling
Aligena[1]	1-2	S	235 m

Generic unit	No. of species (approx.)	Range	Depth and habitat
Americardia	1	S	10 to 65 m
Amiantis	1	S	Int. to 20 m, in sand of exposed beaches
Amygdalum	1	--	75 to 365 m
Anadara	1	S	Int. to 130 m
"Anisodonta"[3]	1	--	20 to 25 m
Anomia	1	S	Int. to 45 m, attached to rocks or other shells
Arcopsis	1	S	Int., among rocks
Argopecten	1	--	Int. to 135 m, free-swimming
Astarte	11-12	N	5 to 230 m
Asthenothaerus	1	S	Int. to 65 m
Atrina	1	S	45 to 90 m
Axinopsida	2	--	10 to 550 m
Bankia	1	--	Int. to 30 m, boring into wood
Barbatia	1	S	Int., under rocks
Barnea	1	--	Int. to 5 m, burrowing in mud
Bathyarca	1-2	S	2,380 to 3,660 m
Bernardina	1	S	Int. to 20 m
Bornia	1	--	Int. to 35 m
Brachidontes	1	S	Int., attached to rocks by byssus
Calyptogena	3	--	500 to 900 m
Cardiomya	8	--	10 to 1,450 m
Chaceia	1	--	Int. to 45 m, boring into rock
Chama	1	--	Int. to 45 m, cemented to rocks or pilings
Chione	3	S	Int. to 45 m, on sand or mud
Chlamydoconcha	1	--	Int. to 25 m
Chlamys	5-6	--	5 to 275 m, free-swimming
Clinocardium	5	--	Int. to 200 m
Compsomyax	1	--	10 to 550 m
Cooperella	1	--	Int. to 200 m, in nest of agglutinated sand
Corbula	2-3	--	Int. to 75 m, among rock rubble
Crassinella	2	--	2 to 35 m
Crassostrea[5]	2	--	Int., in mud in bays; introduced from Japan and Atlantic
Crenella	3-4	--	5 to 455 m
Cryptomya	1	--	Int. to 90 m, in sand or mud
Cumingia	1	--	Int. to 65 m, nestling in crevices or rubble
Cuspidaria	5	--	45 to 2,900 m
Cyathodonta	1	--	25 to 75 m
Cyclocardia	11	--	Int. to 1,830 m
Cyclopecten	6	S	15 to 2,000 m
Cymatioa[2]	1	S	20 m(?)
Cyrtodaria	1	N	About 5 m
Dacrydium	2	--	365 to 2,560 m
Delectopecten	4	--	20 to 2,010 m
Dimya	1-2	S	85 to 185 m, attached to rocks or shells
Diplodonta	3	--	Int. to 220 m, nestling or in sand
Donax	2	--	Int. to 5 m, in sand
Ensis	1	--	Int., in sand
Entodesma	4	--	Int. to 75 m, nestling or attached to substrate by a byssus
Epilucina	1	--	Int. to 75 m, in sand or gravel

Generic unit	No. of species (approx.)	Range	Depth and habitat
Eucrassatella	1	S	20 to 75 m
Felaniella	1	S	Int. to 75 m, in sand or mud
Gari	3	--	Int. to 170 m, in sand or rubble
Gastrochaena[1]	1	S	Int. to 5 m, boring into other shells or calcareous rock
Gemma[5]	1	--	Int. to 10 m, in mud in bays
Glans	1	--	Int. to 90 m, among rocks
Glycymeris	5	--	Int. to 365 m
Gregariella	1	--	35 to 75 m
Grippina	1	S	25 to 35 m
Halicardia	1	--	1,110 to 1,485 m
Halodakra	1-2	S	Int. to 90 m
Heterodonax	1	--	Int., in sand
Hiatella	1-2	--	Int. to 120 m, nestling
Hinnites	1	--	Int. to 55 m, attached to rocks or pilings
Humilaria	1	--	Int. to 50 m, on mud or sand
Huxleyia	1	S	35 to 75 m
Irus	1	--	Int. to 65 m, nestling
Ischadium[5]	1	--	Int., in mud in bays; introduced from Atlantic
Isognomon[3]	1	S	Int. to 5 m, attached to rocks or other shells by a byssus
Kellia	1-2	--	Int. to 65 m, nestling
Laevicardium	2	S	Int. to 135 m, in sand
Lasaea	2-3	--	Int. to 25 m, nestling under rocks
Laternula[5]	1	--	Int., in mud in bays; introduced from Japan
Leporimetis	1	S	Int. to 45 m, in sand
"*Lepton*"[4]	1	--	Not determined
Leptopecten	2	--	Int. to 230 m, attached by a byssus
Lima	5-6	--	Int. to 1,360 m, free-swimming or in transparent nests
Limopsis	4-5	--	55 to 2,000 m
Liocyma	1-2	N	Int. to 220 m
Lithophaga	2	--	Int. to 45 m, boring into rock or other shells
Lucina	2	--	Int. to 45 m, in sand
Lucinoma	2	--	Int. to 550 m
Lyonsia	5	--	Int. to 130 m, in mud of bays
Lyonsiella	1	--	725 m
Lyrodus[5]	1	--	Int. to 10 m, boring in wood
Lyropecten[1]	1	S	5 to 10 m, free-living
Macoma	22	--	Int. to 1,545 m., in sand, silt, or gravel
Mactra	3	--	Int. to 30 m, in sand
Malletia	4-5	--	200 to 2,930 m
Megacrenella	1	--	20 to 400 m
Mercenaria	1	--	Int. to 10 m, in mud of bays
Milneria	2	--	Int. to 25 m, nestling under rocks or on abalones
Miodontiscus	2	--	5 to 130 m
Modiolus	6	--	Int. to 75 m, attached to rocks or in sand
Montacuta	2-3	S	55 to 230 m
Musculus	13	--	Int. to 230 m, nestling in algae or in mud offshore
Mya	6	--	Int. to 50 m, in sandy mud
Myonera	1	--	1,400 to 2,000 m

Generic unit	No. of species (approx.)	Range	Depth and habitat
Mysella	11	--	Int. to 320 m
Mytilimeria	1	--	Int. to 35 m, in the tunicate *Aplidium*
Mytilus	2	--	Int. to 40 m, attached to rocks by a byssus
Neilonella	1	--	2,560 m
Nemocardium	1	--	20 to 180 m
Netastoma	2	--	Int. to 45 m, burrowing into rock
Nucula	10	--	10 to 2,000 m
Nuculana	24-25	--	5 to 3,660 m
Nuttallia	1	--	Int. to 20 m, in sand
Odontogena	1	N	345 m
Orobitella	2-3	--	Int. to 135 m, commensal on crustaceans or annelids
Ostrea	1	--	Int. to 35 m, cemented to hard substrates
Pandora	5	--	5 to 400 m
Panomya	4	N	5 to 300 m
Panopea	1	--	Int. to 20 m, buried about 1 m in sand or mud
Parapholas	1	--	Int. to 45 m, burrowing in rock
Parvilucina	2	--	20 to 245 m
Pecten	2	--	5 to 185 m, free-swimming
Penitella	4	--	Int. to 75 m, boring into rock or shells of the abalone *Haliotis*
Periploma	4	--	20 to 90 m
Petricola	4	--	Int. to 75 m, nestling, boring into hard clay, or on mudflats
Philobrya	1	--	Int. to 75 m, nestling among algae
Pitar	2	--	25 to 185 m
Platyodon	1	--	Int., boring in soft shale
Plectodon	1	S	55 to 110 m
Pododesmus	2	--	Int. to 65 m, attached to rocks or other shells
Policordia	1	--	80 to 2,870 m
Poromya	7	--	75 to 2,930 m
Portlandia	3-4	N	5 to 10 m
Pristes	1	S	Int., in mantle cavity of the chiton *Stenoplax*
Propeamussium	2	--	20 to 700 m
Protothaca	3	--	Int. to 45 m, nestling or in sand or gravel
Psephidia	2-3	--	10 to 90 m, in sandy mud
Pseudochama	2-3	--	Int. to 150 m, cemented to rocks
"*Pseudopythina*"[4]	1	--	Int. to 50 m, commensal on crustaceans
Pteria[3]	1	S	Int. to 10 m, attached to pilings or objects in mud
Raeta[1]	1	S	5 m, in sand
Saxicavella	1	S	45 to 275 m
Saxidomus	2	--	Int. to 35 m, in sand or mud
Semele	5	--	Int. to 190 m, nestling or in sand
Septifer	1	--	Int., attached to rocks by a byssus
Serripes	2	N	Int. to 135 m
Siliqua	5	--	Int. to 60 m, in sand
Solemya	3	--	7 to 3,290 m
Solen	2	--	Int. to 75 m
Sphenia	2	--	Int. to 65 m, nestling
Spinula[2]	1	--	4,100-6,100 m
Spisula	6	--	Int. to 100 m, in sand of exposed beaches or bays

Generic unit	No. of species (approx.)	Range	Depth and habitat
Sportella	1	--	Not known
Tagelus	3	--	Int., in sand or mud of bays
Tapes[5]	1	--	Int. to 10 m, on mudflats; introduced from Japan
Tellina	7	--	Int. to 440 m, in silt to sand
Teredo[5]?	1	--	Int. to 10 m, boring into wood
Theora[5]	1	S	About 5 m, in mud of bays; introduced from Japan
Thracia	6	--	Int. to 135 m, nestling
Thyasira	7	--	5 to 2,010 m
Tindaria	8	--	365 to 2,560 m
Tivela	1	--	Int. to 5 m, in sand of exposed beaches
Tomburchus	1	S	90 to 235 m
Trachycardium	1	S	Int. to 120 m, in sand
Transennella	1-2	--	Int. to 35 m, in fine sand
Tresus	2	--	Int. to 35 m, in sand
Turtonia	2	--	Int. to 20 m, nestling in algae holdfasts
Ventricolaria	1	--	10 to 75 m
Verticordia	2	S	35 to 1,200 m
Vesicomya	4	--	550 to 3,110 m
Xylophaga	2	--	20 to 805 m, burrowing into wood
Xyloredo	1	S	2,075 m, in mud
Yoldia	12-13	--	Int. to 2,000 m
Yoldiella	7-8	--	10 to 1,830 m
Zirfaea	1	--	Int. to 10 m, burrowing in mud or shale

POLYPLACOPHORA

[All genera clinging to rocks or other shells]

Generic unit	No. of species (approx.)	Range	Depth and habitat
Acanthochitona	1	--	Low int. to 20 m
Amicula	3	N	Int. to 135 m
Basiliochiton	3	--	Int. to 135 m
Callistochiton	3	--	Low int. to 65 m
Chaetopleura	1	--	Low int. to 25 m
Cryptochiton	1	--	Low int. to 20 m
Cyanoplax	4	--	High int. to 80 m
Dendrochiton	2	--	Low int. to 90 m
Hanleya	1	N?	45 to 185 m
Ischnochiton	12-13	--	Low int. to 2,000 m
Katharina	1	--	High to low int.
Lepidochitona	1-2	--	Int. to 300 m
Lepidozona	12	--	Low int. to 730 m
Leptochiton	10	--	Int. to 3,660 m
Mopalia	16-17	--	High int. to 200 m
Nuttallina	3	--	High to low int.
Oldroydia	1	--	Low int. to 45 m
Placiphorella	4	--	Low int. to 2,000 m
Schizoplax	1	N	Int. to 25 m
Spongioradsia	1	N	Low int. to 20 m
Stenoplax	4	--	Low int. to 50 m
Tonicella	5	--	Low int. to 300 m

Generic unit	No. of species (approx.)	Range	Depth and habitat
		APLACOPHORA	
Alexandromenia	1	S	1,105 to 2,470 m, on mud
Chaetoderma	18	--	20 to 1,595 m, on mud or sand
Dondersia	1	S	About 40 m
Dorymenia	1	S	550 to 1,165 m, probably on hydroids
Heathia	1	S	915 to 990 m, possibly on sea pens
Limifossor	2	--	145 to 825 m, on silt
Nematomenia	1	N	880 m, on hydroids
Pachymenia	1	S	4,020 to 4,075 m
Platymenia	1	S	730 m, on mud
Prochaetoderma	1	S	535 to 830 m, on mud
Proneomenia	1	N	200 m
		SCAPHOPODA [All genera shallowly buried in sea floor]	
Cadulus	7	--	7 to 3,050 m
Dentalium	10	--	5 to 2,320 m
Fustiaria	1	--	2 to 1,900 m
Siphonodentalium	1	--	5 to 365 m
		CEPHALOPODA	
Abraliopsis	1	S	No precise records on this genus
Alloposus	1	S	No precise records on this genus
Argonauta	2	S	No precise records on this genus
Bathyteuthis	1	S	800 to 1,300 m
Benthoctopus	2	N	200 to 1,000 m
Berryteuthis	2		90 to 1,830 m
Chiroteuthis	1	--	150 to 2,130 m
Cranchia	1	--	1 to 3,500 m
Crystalloteuthis	1	N	105 to 4,800 m
Dosidicus	1	--	No precise records on this genus
Eledonella	1	S	No precise records on this genus
Galiteuthis	2	--	100 to 4,000 m
Gonatopsis	1	--	About 365 m
Gonatus	5-7	--	60 to 1,830 m
Grimalditeuthis	1	S	No precise records on this genus
Grimpoteuthis	1	S	350 to 535 m
Helicocranchia	1	S	About 640 m
Histioteuthis	3	--	100 to 1,300 m
Japetella	1-2	--	2,000 to 4,075 m
Leachia	1	S	1 to 1,100 m
Loligo	1	--	1 to 185 m
Mastigoteuthis	1	S	No precise records on this genus
Moroteuthis	1	--	1 to 550 m
Octopoteuthis	1	--	1 to 2,000 m
Octopus	7-8	--	Int. to 1,900 m
Ocythoe	1	S	No precise records on this genus
Ommastrephes	1	--	1 to 1,490 m

Generic unit	*No. of species (approx.)*	*Range*	*Depth and habitat*
Onychoteuthis	1	--	1 to 4,000 m
Opisthoteuthis	1	--	125 to 915 m
Pterygioteuthis	2	S	10 to 2,500 m
Pyroteuthis	1	S	No precise records on this genus
Rossia	1	--	15 to 185 m
Symplectoteuthis	1	S	1 to 1,300 m
Taonius	1	--	200 to 1,000 m
Todarodes	1	N	500 m
Valbyteuthis	2	S	1 to 2,330 m
Vampyroteuthis	1	--	510 to 4,000 m

IDENTIFICATION OF FIGURES

As noted in the Introduction, this is a list of the species that are figured in the keys as representatives of their respective genera or subgenera; subgenera, of course, are cited within conventional parentheses. Many of the figures have been deliberately made composite or generalized, and for these no specific name is cited. It would be impractical, if not impossible, to supply the sources of the figures themselves; but wherever a species is given, the author of that taxon and the date of the original published description are given in lieu of a bibliographic reference (i.e., these publications are not to be found in the Bibliography proper).

Arrangement is alphabetical by class. A single illustration has been used at some places in the keys for two or more taxa separable only on soft parts; for these, the listings are cross-referenced. Rarely, the shell illustrated may not be of Western North American origin, though morphologically characteristic; such figures are indicated by parenthetical information, e.g. "(Panamic)." Finally, the use of parentheses around the author's name for some species is a convention required under the International Code to show that the species was originally assigned by the author to some genus other than the one now in use and that it has since been transferred—a new combination.

GASTROPODA

Acanthina spirata (Blainville, 1832)
Aclis (?Aclis) californica Bartsch, 1927
A. (Graphis) shepardiana (Dall, 1919)
A. (Schwengelia) occidentalis (Hemphill, 1894)
Acmaea mitra Rathke, 1833
Acteocina, generalized
Acteon traski Stearns, 1898
Admete couthouyi (Jay, 1839)
Aesopus chrysalloideus (Carpenter, 1864)
Aglaja diomedea Bergh, 1894
Alaba jeannettae Bartsch, 1910
Alabina monicensis Bartsch, 1911 (Pleistocene)
Alvinia, generalized
Amauropsis purpurea Dall, 1871 [?=*A. islandica* (Gmelin, 1791)]
Amphissa versicolor Dall, 1871
Amphithalamus inclusus Carpenter, 1865
Anabathron muriei Bartsch & Rehder, 1939
Ancistrolepis eucosmius (Dall, 1891)
Antiplanes (Antiplanes) perversa (Gabb, 1865)
A. (Rectiplanes) santarosana (Dall, 1902)
A. (Rectisulcus) strongi (Arnold, 1903)
Aplysia, generalized
Aquilonaria turneri Dall, 1886
Arctomelon stearnsii (Dall, 1872)
Assiminea californica (Tryon, 1865)
Astraea undosa (Wood, 1828)
Atlanta, generalized
Atys, generalized
Austrotrophon catalinensis (Oldroyd, 1927)
Balcis compacta (Carpenter, 1864)
Barleeia (Barleeia) sanjuanensis Bartsch, 1920

B. (Pseudodiala), generalized
Bathybembix (Bathybembix) bairdii (Dall, 1889)
B. (Cidarina) cidaris (Carpenter, 1864)
Batillaria zonalis (Bruguière, 1792)
Bellaspira grippi (Dall, 1908)
Beringius, generalized
Berthella californica (Dall, 1900)
Berthellina: see *Berthella*
Bittium, generalized
Borsonella, generalized
Buccinum strigillatum Dall, 1891
Bulbus apertus (Lovén, 1847) [?=*B. fragilis* (Leach, 1819)]
Bulla, generalized
Bursa californica (Hinds, 1843)
Busycon (Busycotypus) canaliculatum (Linnaeus, 1758)
Caecum, generalized
Calinaticina oldroydii (Dall, 1897)
Calliostoma, generalized
Calliotropis carlotta (Dall, 1902)
Calyptraea fastigiata Gould, 1846
Cancellaria crawfordiana Dall, 1891
Capulus californicus Dall, 1900
Cardiapoda placenta (Lesson, 1830)
Carinaria, generalized
Carinoturris adrastia (Dall, 1919)
Cavolinia tridentata (Forskål, 1773)
Cecina manchurica A. Adams, 1861
Ceratostoma nuttalli (Conrad, 1837)
Cerithidea californica (Haldeman, 1840)
Cerithiopsis, generalized
Chelidonura inermis (Cooper, 1862)
Choristes carpenteri Dall, 1896
Cingula aleutica (Dall, 1887)
Clathurella (Clathurella), generalized
C. (Crockerella) crystallina (Gabb, 1865)
Clio pyramidata Linnaeus, 1767
Cocculina, generalized
Collisella, generalized
Colus trophius (Dall, 1919)
Conus californicus Reeve, 1844, *ex* Hinds MS
Crassispira, s.l., generalized
C. (Burchia) semiinflata (Grant & Gale, 1931)
Crepidula nummaria Gould, 1846
Crepipatella lingulata (Gould, 1846)
Creseis virgula (Rang, 1828)
Crucibulum spinosum (Sowerby, 1824)
Cryptobranchia, generalized
Cryptogemma, generalized
Cuvierina columnella (Rang, 1827)
Cyclostremella californica Bartsch, 1907
Cyclostremiscus cerrosensis (Bartsch, 1907)
Cylichna, generalized
Cymakra intermedia (Arnold, 1903)
Cymatium tremperi Dall, 1907
Cypraea spadicea Swainson, 1823
Cystiscus, generalized
Cytharella victoriana (Dall, 1897)
Daphnella, generalized
Dendropoma lituella (Mörch, 1861)
Diacria quadridentata (Blainville, 1821)
Diaphana, generalized
Diastoma fastigiata (Carpenter, 1864)
Diodora aspera (Rathke, 1833)
Elachisina grippi Dall, 1918
Elaeocyma, generalized
Elephantanellum heptagonum (Carpenter, 1857) (Panamic)
Engina strongi Pilsbry & Lowe, 1932
Episcynia devexa Keen, 1946
Epitonium sawinae (Dall, 1903)
Erato, generalized
Eulima, generalized
Eupleura grippi Dall, 1911
Exilioidea, generalized
Fartulum, generalized
Fissurella volcano Reeve, 1849
Fissurisepta soyoae Kuroda, 1951 (Japan)
Forreria belcheri (Hinds, 1844)
Fusinus arnoldi (Cossmann, 1903)
Fusitriton oregonensis (Redfield, 1848)
Gastropteron pacificum Bergh, 1893
Granula polita (Carpenter, 1857)
Granulina margaritula (Carpenter, 1857)
Haliella, generalized
Haliotis, generalized
Halistylus pupoideus (Carpenter, 1864)
Haloconcha reflexa (Dall, 1884)
Haminoea, generalized
Hipponix tumens Carpenter, 1864
Homalopoma, generalized
Hyalocylis striata (Rang, 1828)
Iselica fenestrata (Carpenter, 1864)
Janthina, generalized
Kelletia kelletii (Forbes, 1852)
Lacuna porrecta Carpenter, 1864
Lamellaria, generalized
Latiaxis oldroydi (I. Oldroyd, 1929)
Lepeta, generalized
Leptogyra alaskana Bartsch, 1910
Leucosyrinx (Aforia) circinata (Dall, 1873)
Limacina helicina (Phipps, 1774)
Liomesus nux Dall, 1877
Liotia fenestrata Carpenter, 1864
Lirularia succincta (Carpenter, 1864)
Litiopa melanostoma Rang, 1829

Littorina (Littorina), generalized
L. (Algamorda) newcombiana (Hemphill, 1876)
Lottia gigantea Sowerby, 1833
Lucapinella callomarginata (Dall, 1871)
Macrarene californica (Dall, 1908)
Macromphalina californica Dall, 1903
Macron lividus (A. Adams, 1855)
Mangelia, s.l., generalized
Margarites (Margarites) marginatus Dall, 1919
M. (Pupillaria) pupillus (Gould, 1849)
Maxwellia gemma (Sowerby, 1879)
Megasurcula tremperiana (Dall, 1911)
Megatebennus bimaculatus (Dall, 1871)
Megathura crenulata (Sowerby, 1825)
Melampus olivaceus Carpenter, 1857
Metaxia convexa (Carpenter, 1857) (Panamic)
Metzgeria californica Dall, 1903
Micranellum, generalized
Microglyphis breviculus (Dall, 1902)
Mistostigma albidum (Carpenter, 1864)
Mitra idae Melvill, 1893
Mitrella tuberosa (Carpenter, 1864)
Mitromorpha, generalized
Moelleria quadrae Dall, 1897
Mohnia frielei Dall, 1891
Morrisonella pacifica (Dall, 1908)
Murexiella santarosana (Dall, 1905)
Nannoteretispira kelseyi (Bartsch, 1911)
Nassarina penicillata (Carpenter, 1864)
Nassarius (Demondia) rhinetes Berry, 1953
N. (Ilyanassa) obsoletus (Say, 1822)
Natica (Cryptonatica), generalized
Neadmete circumcincta (Dall, 1873)
Neptunea, generalized
Neverita reclusiana (Deshayes, 1839)
Niso hipolitensis Bartsch, 1917
Norrisia norrisii (Sowerby, 1838)
Notoacmea: see *Collisella*
Nucella: upper figure, *N. canaliculata* (Duclos, 1832); lower figure, *N. lamellosa* (Gmelin, 1791)
Ocenebra interfossa Carpenter, 1864
Odostomia, generalized
O. (Chrysallida) montereyensis Dall & Bartsch, 1907
Oenopota, generalized
Olivella biplicata (Sowerby, 1825)
Onchidiopsis, generalized (based on Atlantic specimens)
Opalia, generalized
Ophiodermella, generalized
Oxygyrus, generalized (Atlantic)
Parviturbo acuticostatus (Carpenter, 1864)
Pedicularia californica Newcomb, 1864
Pedipes unisulcatus Cooper, 1866
Peracle bispinosa Pelseneer, 1888
Peristichia pedroana (Dall & Bartsch, 1909)
Petaloconchus: smaller figure, *P. montereyensis* Dall, 1919; larger figure, *P. macrophragma* Carpenter, 1857 (the latter Panamic, showing internal lamina)
Philine, generalized
Phyllaplysia taylori Dall, 1900
Phytia, generalized
Piliscus commodus (Middendorff, 1851)
Pleurobranchus: see *Berthella*
"*Pleurotomella*," generalized
Plicifusus brunneus (Dall, 1877)
Polinices, generalized
Protatlanta souleyeti (E. A. Smith, 1888)
Pseudomelatoma penicillata (Carpenter, 1864)
Pteropurpura (Pteropurpura) alba (Berry, 1908)
P. (Shaskyus) festiva (Hinds, 1844)
Pterynotus (Pterochelus) phillipsi Vokes, 1966
Ptychatractus occidentalis (Stearns, 1871)
Puncturella cucullata Gould, 1846
Pyramidella adamsi Carpenter, 1864
Pyrulofusus harpa (Mörch, 1858)
Retusa, generalized
Rhodopetoma rhodope (Dall, 1919)
Rictaxis punctocoelata (Carpenter, 1864)
Rimula californiana Berry, 1964
Rissoella, generalized
Rissoina californica Bartsch, 1915
Roperia poulsoni (Carpenter, 1865)
Sabinella ptilocrinicola (Bartsch, 1907)
Scaphander, generalized
Scelidotoma bella (Gabb, 1865)
Scissilabra dalli Bartsch, 1907
Scissurella crispata Fleming, 1832
Searlesia dira (Reeve, 1846)
Seguenzia, generalized
Seila assimilata (C. B. Adams, 1852) (Panamic)
Serpulorbis squamigerus (Carpenter, 1857)
Simnia loebbeckeana (Weinkauff, 1881)
Sinezona rimuloides (Carpenter, 1865)
Sinum scopulosum (Conrad, 1849)
Siphonaria brannani Stearns, 1872
Skeneopsis alaskana Dall, 1919
Solariella peramabilis Carpenter, 1864
Solariorbis arnoldi Bartsch, 1927
Styliola, generalized

Suavodrillia kennicotti (Dall, 1871)
Sulcoretusa, generalized
Sulcosinus taphrius (Dall, 1891)
Surculina cortezi (Dall, 1908)
Tachyrynchus lacteolus (Carpenter, 1864)
Tegula funebralis (A. Adams, 1854)
Teinostoma politum A. Adams, 1851 (Panamic; type of genus)
Terebra, generalized
Torellia ammonia Dall, 1919
Trachypollia lugubris (C. B. Adams, 1852)
Trichotropis, generalized
Tricolia, generalized
Trimusculus reticulatus (Sowerby, 1835)
Triphora, generalized
Trivia californiana (Gray, 1828)
Trophonopsis lasia (Dall, 1919)
Truncatella californica Pfeiffer, 1857
Turbonilla, s.s., generalized
T. (Pyrgolampros) pedroana Dall & Bartsch, 1903
Turcica caffea (Gabb, 1865)
Turritella cooperi Carpenter, 1864
Turritellopsis, generalized
Tylodina fungina Gabb, 1865
Urosalpinx cinerea (Say, 1822)
Velutina, generalized
Vermetus indentatus (Carpenter, 1857) (Panamic; similar to *V. compactus* (Carpenter, 1864) from Puget Sound)
Vermicularia eburnea (Reeve, 1842) (Panamic)
Vitrinella oldroydi Bartsch, 1907
Vitrinorbis diegensis (Bartsch, 1907)
Volutharpa ampullacea (Middendorff, 1848)
Volutomitra alaskana Dall, 1902
Volutopsius simplex Dall, 1907
Volvarina taeniolata Mörch, 1860
Volvulella cylindrica (Carpenter, 1864)
Williamia peltoides (Carpenter, 1864)
?*Woodbridgea williamsi* Berry, 1953 (Panamic)

PELECYPODA

Acila castrensis (Hinds, 1843)
Adontorhina cyclia Berry, 1947
Adula californiensis (Philippi, 1847)
Aligena cerritensis Arnold, 1903
Americardia biangulata (Broderip & Sowerby, 1829)
Amiantis callosa (Conrad, 1837)
Amygdalum pallidulum (Dall, 1916)
Anadara multicostata (Sowerby, 1833)
Anomia peruviana Orbigny, 1846
Arcopsis solida (Sowerby, 1833)
Argopecten aequisulcatus (Carpenter, 1864)
Astarte, generalized
Asthenothaerus villosior Carpenter, 1864
Atrina oldroydii Dall, 1901
Axinopsida serricata (Carpenter, 1864)
Bankia setacea Tryon, 1863
Barbatia, generalized
Barnea subtruncata (Sowerby, 1834)
Bathyarca orbiculata (Dall, 1881)
Bernardina bakeri Dall, 1910
Bornia retifera Dall, 1899
Brachidontes adamsianus (Dunker, 1857)
Calyptogena pacifica Dall, 1891
Cardiomya, generalized
Chaceia ovoidea (Gould, 1851)
Chama: "*C. pellucida* Sowerby, 1834," auctt., not of Sowerby
Chione (Chione) undatella (Sowerby, 1835)
C. (Chionista) fluctifraga (Sowerby, 1853)
Chlamydoconcha orcutti Dall, 1884
Chlamys, generalized
Clinocardium nuttallii (Conrad, 1837)
Compsomyax subdiaphana (Carpenter, 1864)
Cooperella subdiaphana (Carpenter, 1864)
Corbula, generalized
Crassinella, generalized
Crassostrea, generalized
Crenella divaricata (Orbigny, 1846)
Cryptomya californica (Conrad, 1837)
Cumingia californica Conrad, 1837
Cuspidaria, generalized
Cyathodonta dubiosa Dall, 1915
Cyclocardia, generalized
Cyclopecten, generalized
Cyrtodaria kurriana Dunker, 1861
Dacrydium pacificum Dall, 1916
Delectopecten, generalized
Dimya californiana Berry, 1936
Diplodonta, generalized
Donax gouldii Dall, 1921
Ensis myrae Berry, 1953
Entodesma saxicola (Baird, 1863)
Epilucina californica (Conrad, 1837)
Eucrassatella, generalized
Felaniella sericata (Reeve, 1850)

Gari californica (Conrad, 1849)
Gemma gemma (Totten, 1834)
Glans subquadrata (Carpenter, 1864)
Glycymeris subobsoleta (Carpenter, 1864)
Gregariella chenui (Récluz, 1842)
Grippina californica Dall, 1912
Halodakra, generalized
Heterodonax pacificus (Conrad, 1837)
Hiatella, generalized
Hinnites giganteus (Gray, 1825)
Humilaria kennerleyi (Reeve, 1863)
Huxleyia munita (Dall, 1898)
Irus lamellifer (Conrad, 1837)
Ischadium (Geukensia) demissum (Dillwyn, 1817)
Kellia laperousii (Deshayes, 1839)
Laevicardium substriatum (Conrad, 1837)
Lasaea cistula Keen, 1938
Laternula limicola (Reeve, 1863); hinge from *L. anatina* (Linnaeus, 1758), from East Indies
Leporimetis obesa (Deshayes, 1855)
"Lepton" meroeum Carpenter, 1864
Leptopecten, generalized
Lima hemphilli Hertlein & Strong, 1946
Limopsis, generalized
Liocyma fluctuosa (Gould, 1841)
Lithophaga attenuata rogersi Berry, 1957
Lucina (Here) richthofeni Gabb, 1866
L. (Lucinisca) nuttalli Conrad, 1837
Lucinoma annulata (Reeve, 1850)
Lyonsia, generalized
Lyonsiella quaylei Bernard, 1969
Lyrodus pedicellatus (Quatrefages, 1849)
Lyropecten subnodosus (Sowerby, 1835)
Macoma, generalized
Mactra, generalized
Malletia, generalized
Megacrenella columbiana (Dall, 1897)
Mercenaria, generalized
Milneria minima (Dall, 1871)
Miodontiscus prolongatus (Carpenter, 1864)
Modiolus, generalized
Montacuta balliana (Dall, 1916)
Musculus, generalized
Mya arenaria Linnaeus, 1758
Myonera tillamookensis Dall, 1916
Mysella, generalized
Mytilimeria nuttalli Conrad, 1837
Mytilus edulis Linnaeus, 1758
Neilonella brunnea (Dall, 1916)
Nemocardium centifilosum (Carpenter, 1864)
Netastoma rostrata (Valenciennes, 1846)
Nucula, generalized
Nuculana, generalized
Nuttallia nuttallii (Conrad, 1837)
Odontogena borealis (Cowan, 1964)
Orobitella, generalized
Ostrea lurida Carpenter, 1864
Pandora, generalized
Panomya, generalized
Panopea generosa Gould, 1850
Parapholas californica (Conrad, 1837)
Parvilucina tenuisculpta (Carpenter, 1864)
Pecten diegensis Dall, 1898
Penitella penita (Conrad, 1837)
Periploma planiusculum Sowerby, 1834
Petricola (Petricola) carditoides (Conrad, 1837)
P. (Petricolaria) pholadiformis Lamarck, 1818
Philobrya setosa (Carpenter, 1864)
Pitar, generalized
Platyodon cancellatus (Conrad, 1837)
Plectodon scaber Carpenter, 1864
Pododesmus (macrochisma?) cepio (Gray, 1850)
Policordia alaskana (Dall, 1895)
Poromya trosti Hertlein & Strong, 1937
Portlandia arctica (Gray, 1824)
Pristes oblongus Carpenter, 1864
Propeamussium alaskense (Dall, 1871)
Protothaca staminea (Conrad, 1837)
Psephidia lordi (Baird, 1863)
Pseudochama exogyra (Conrad, 1837)
"Pseudopythina" compressa (Dall, 1899)
Saxicavella pacifica Dall, 1916
Saxidomus nuttalli Conrad, 1837
Semele decisa (Conrad, 1837)
Septifer bifurcatus (Conrad, 1837)
Serripes groenlandicus (Bruguière, 1789)
Siliqua patula (Dixon, 1789)
Solemya, generalized
Solen sicarius Gould, 1850
Sphenia, generalized
Spisula, generalized
Sportella: (a) *S. californica* Dall, 1899; (b) *S. stearnsii* Dall, 1899 (Gulf of California)
Tagelus californianus (Conrad, 1837)
Tapes japonica Deshayes, 1853
Tellina, s.l.: *T. modesta* (Carpenter, 1864)
Tellina (Megangulus) lutea alternidentata Broderip & Sowerby, 1829
Teredo, generalized
Theora lubrica Gould, 1861
Thracia trapezoides Conrad, 1849

Thyasira disjuncta (Gabb, 1866)
Tindaria, generalized
Tivela stultorum (Mawe, 1823)
Tomburchus redondoensis (T. Burch, 1941)
Trachycardium quadragenarium (Conrad, 1837)
Transennella tantilla (Gould, 1853)
Tresus nuttallii (Conrad, 1837)
Turtonia, generalized
Ventricolaria fordii (Yates, 1890)
Verticordia ornata (Orbigny, 1846)
Vesicomya lepta (Dall, 1896)
Xylophaga washingtona Bartsch, 1921
Xyloredo: see *Xylophaga*
Yoldia, generalized
Yoldiella, generalized
Zirfaea pilsbryi Lowe, 1931

POLYPLACOPHORA

Acanthochitona diegoensis (Pilsbry, 1893)
Amicula vestita (Broderip & Sowerby, 1829)
Basiliochiton, generalized
Callistochiton palmulatus Dall, 1879
Chaetopleura gemma Dall, 1879
Cryptochiton stelleri (Middendorff, 1846)
Cyanoplax hartwegii (Carpenter, 1856)
Dendrochiton, generalized
Hanleya hanleyi (Bean, 1844)
Ischnochiton, generalized
Katharina tunicata (Wood, 1815)
Lepidochitona, generalized
Lepidozona cooperi (Dall, 1879)
Leptochiton, generalized
Mopalia ciliata (Sowerby, 1840)
Nuttallina californica (Reeve, 1847)
Oldroydia percrassa (Dall, 1894)
Placiphorella velata Dall, 1879
Schizoplax brandtii (Middendorff, 1846)
Spongioradsia aleutica (Dall, 1878)
Stenoplax conspicuus (Pilsbry, 1892, *ex* Carpenter MS)
Tonicella lineata (Wood, 1815)

APLACOPHORA

Chaetoderma montereyense Heath, 1911
Dorymenia acuta Heath, 1911

SCAPHOPODA

Cadulus, generalized
Dentalium, generalized
Fustiaria dalli (Pilsbry & Sharp, 1897)
Siphonodentalium quadrifissatum (Pilsbry & Sharp, 1898)

CEPHALOPODA

Argonauta nouryi Lorois, 1852
Beaks and "pens," generalized

GLOSSARY

All the morphological, technical, and nomenclatural terms used in the foregoing sections are defined in this Glossary. Although many of the terms used in the keys can be found in an ordinary dictionary, some of them have acquired special meanings, which we attempt to define. Additional terms and more extensive discussions are given in Arnold (1966), Cox in Moore et al. (1960, 1969-71), and the *International Code of Zoological Nomenclature* (1961, 1964).

A. Terms Used in Describing Specimens and Habitats

ABYSSAL: Associated with the abyssal zone, i.e., the floor of the deeper ocean from 2,000 to 6,000 meters.

ACCESSORY PLATE: A secondary calcareous or corneous structure formed in some pelecypods (e.g., Pholadidae) to protect the soft parts.

ACCESSORY SCAR: In pelecypods, a small, mostly circular impression on the inside of the shell made by other than the adductor and pallial muscles.

ACUTE: Sharply angled.

ADDUCTOR: In pelecypods, a large muscle that pulls the two valves together.

ADDUCTOR SCAR: A differentiated area on the interior of a pelecypod shell that marks the attachment point of an adductor and is generally striated, depressed, or outlined.

ADHERENT: Closely attached.

ANGULATE: Formed with corners; angled.

ANNULATION OF ANNULAR RING: A growth increment in a tubular shell, marked by regular constrictions (e.g., *Caecum*).

ANTERIOR: Forward or head end, defined in the various molluscan classes as indicated on the illustrations at the beginnings of the several keys. *See also* Posterior; Dorsal; Ventral.

ANTERIOR CANAL: A tubular or troughlike extension at the anterior end of a gastropod aperture, enclosing the inhalant siphon.

APERTURE: In gastropods, the major opening of a shell, bounded by the shell's last-formed margin.

APEX: The initial point from which a gastropod or scaphopod shell is built, generally pointed.

APICAL CAVITY: A space under the apex of patellate gastropods.

APICAL ORIFICE: An opening at the apex.

APICAL PLUG: A special calcareous filling in the apex of certain gastropods (e.g., *Fartulum*).

APOPHYSIS: A projecting structure such as that serving as a muscle attachment in the Pholadidae.

APPRESSED: With whorls overlapping, so that their outer surfaces converge gradually. Also given (in other works) as Adpressed.

ARTICULAMENTUM: In chitons, the middle shell layer.

ATTACHMENT SCAR: Any impression left on a molluscan shell by the attachment of a soft part (e.g., mantle, muscle, or foot).

AURICLE: In some pelecypods (e.g., scallops), either of the earlike projections near the hinge.

AXIAL: Roughly parallel to the axis of coiling in a gastropod shell.

BAND: A strip of shell material differentiated by color or construction from the shell on either side of it.

BANDING: Color marking in continuous stripes.

BASAL: Associated with the base of a gastropod shell.

BASAL FOLD: A fold near the anterior end of the columella on a gastropod shell.

BASAL PLATE: A segment of the ribbon to which radula teeth are attached (e.g., in *Collisella*).

BASE: In coiled gastropods, the area below the periphery of the body whorl, excluding the aperture; in uncoiled or limpet-like shells, the rim of the aperture.

BATHYMETRIC: Pertaining to the measurement of depth in bodies of water; also, pertaining to the distribution of organisms over various depths.

BEADED: Sculptured so as to resemble beads or strings of beads.

BEAK: The small tip of a pelecypod shell, near the hinge; also, any spoutlike elongation of a shell (e.g., *Cuspidaria*). *See also* Umbo.

BENTHONIC: Living on the sea bottom. Also given as Benthic.

BICONIC: Having a diamond-shaped outline; especially used in reference to gastropod shells having a spire of about the same size and shape as the body whorl.

BIFID: Divided into two parts by a groove; applied especially to the hinge teeth of pelecypods.

BIFURCATION: A division into two branches, especially in shell sculpture.

BODY WHORL: In a coiled gastropod, the terminal and largest whorl of the shell; it is the section closest to the aperture and encloses most of the body of the animal.

BRIDGED: Extending from one side to the other; partly covered.

BUCCAL: Pertaining to the organs of the mouth area in gastropods, especially to the bulging flexible mass that supports the radula.

BULBOUS: Bulging or globular.

BULLOID: Bubble-shaped; in the shape of a *Bulla* shell.

BUTTRESS: A shell-strengthening structure—e.g., a supporting laminar costa in gastropods, or a support for part of the hinge in pelecypods.

BYSSUS: A bundle of tough conchiolin strands secreted by the foot of some pelecypods, passing out the anterior end of the shell and used to secure the animal to a substrate.

CALCAREOUS: Composed mostly of calcium carbonate.

CALCIFIED: With the conchiolin matrix partially or entirely reinforced by calcium carbonate intercalations.

CALLUM: Shell material filling the pedal gape in certain mature pholad pelecypods.

CALLUS: A thickened layer of shelly substance, especially around the aperture in gastropods. Also called (in other works) the Inductura.

CANAL: In gastropods, a narrow notch or tubular extension of the aperture, enclosing a siphon.

CANCELLATE: Having a pattern of sculptural lines that intersect at right angles; reticulate. *See also* Decussate.

CARDINAL: In pelecypods, situated more or less in the central part of the hinge area directly below the beaks.

CARDINAL TEETH: In heterodont pelecypods, the teeth radiating on the hinge plate immediately below the beaks.

CARINA: A prominent, sharp-edged ridge; same as keel.

CARINATE: Having a carina.

CARTILAGE: A translucent, elastic tissue inside the hinge that supplements the external ligament in binding the two valves of a pelecypod together, attached to the resilifers and perhaps reinforced by a calcareous coating, the lithodesma. The term "internal ligament" is preferred by modern authors.

CARTILAGINOUS: Having a flexible or horny texture, as contrasted with calcareous (shelly).

CEMENTATION: Fixation to the substrate in sessile mollusks.

CHAFFY: Covered with a roughened surface, as by scales.

CHANNELED: With a deep groove, or sculptured with a series of grooves.

CHINK: A long, narrow cleft, as in the umbilicus of *Lacuna*. *See also* Gape; Rimate.

CHITIN: A tough, horny substance that forms the exoskeleton of arthropods: in mollusks it sometimes occurs in the radula of a gastropod; however, most horny material in mollusks is conchiolin, not chitin. Thus the term, though often used in the literature, is to be avoided unless a chemical analysis of the ligament, operculum, or other structure is made to confirm the presence of chitin.

CHONDROPHORE: A large, spoon-shaped resilifer. *See also* Fossette.

COALESCED: Fused or merged together.

COLLAR: A raised lip bordering a suture.

COLUMELLA: A pillar surrounding the axis of coiling in most gastropods, formed by the inner walls of the whorls.

COMMENSAL: Living with another animal, but not deriving nourishment directly from the host. *See also* Parasitic.

COMPRESSED: Flattened; with reduced thickness.

CONCENTRIC SCULPTURE: Sculpture congruent with or roughly paralleling the ventral margin of a pelecypod shell, and contrasted with radial sculpture; in the same direction but heavier than growth lines.

CONCHIOLIN: A proteinaceous material that makes up the periostracum of a shell, and also forms the organic matrix for calcareous parts of the shell.

CONSTRICTION: A narrowing or "waist" in a shell structure.

CORDS: Round-topped, moderately coarse spiral or axial features of shell sculpture.

CORNEOUS: Horny in texture.

CORRUGATED: Folded or ridged; broadly and heavily sculptured.

COSTA (pl. COSTAE): A riblike sculptural element on a shell surface. *See also* Rib.

CRENATE OR CRENULATE: Having a regularly notched edge.

CUTICLE: *See* Epidermis.

DECK: A septum or transverse plate of shelly material, as in *Crepidula. See also* Shelf.

DECUSSATE: Having a latticed surface formed by the intersection of fine ribs, not necessarily at right angles. *See also* Cancellate.

DENTATE: Having conspicuous projections along a margin. *See* Denticles.

DENTICLES: Small projections around the margin of a gastropod aperture or the margin of a pelecypod valve, especially near the hinge. (Not to be confused with true interlocking hinge teeth.)

DENTICULATE: Having denticles.

DENTITION: Tooth structure: referring in pelecypods to the hinge teeth, in gastropods usually to the elements of the radula.

DEPRESSED: Low in proportion to diameter.

DEXTRAL: In gastropods, coiling so that the aperture is at the right when the shell is seen in ventral view with spire uppermost. *See also* Sinistral.

DIAMETER: In gastropods, the greatest width of the shell; in pelecypods, a measure of shell convexity.

DIVARICATE: Diverging sculpture, often forming chevrons.

DORSAL: In gastropods, lying on the side of the shell opposite the aperture; in bivalves, lying on the surface or margin nearest the hinge; in chitons, lying on the body surface that bears the eight shell valves. *See also* Ventral; Anterior; Posterior.

EAR: A triangular, lateral prolongation of the dorsal margin of a bivalve shell.

EDENTATE: Without hinge teeth.

ELONGATE: Extended; considerably longer in one dimension than another.

EMBAYMENT: Curved portion of a pallial sinus in pelecypods.

ENTIRE: Oval or rounded and smoothly arched, uninterrupted by any re-entrant curve, sinus, or notch.

EPIDERMIS: The outermost layer of the molluscan body, not associated with the shell; cuticle; integument. (Often erroneously applied to the periostracum, the outer layer of the shell.)

EQUIVALVE: With valves equal in size and shape.

ERODED: Having parts of the surface worn away.

ESCUTCHEON: A somewhat lozenge-shaped differentiated dorsal area of the pelecypod shell, extending posteriorly from the beaks and often bordered by a ridge.

EXPANDED: With dimensions proportionately greater than those of preceding parts of the shell.

EXTERNAL LIGAMENT: That portion of the ligament visible when the valves of a pelecypod are closed. *See* Ligament.

FASCIOLE: In gastropod shells, a prominent spiral band formed by successive margins of a canal or notch.

FISSURE: A narrow crack, slot, or deep groove.

FLARING: Opening outward, widening, often used with reference to the aperture of a gastropod shell.

FLEXED: Folded or warped.

FLEXURE: A progressive folding or warping of one or both valves in a pelecypod; more generally, a bending or angulation.

FOLD: A spirally wound ridge on the columellar wall of a gastropod shell.

FOOT: A characteristic muscular organ of locomotion present in most mollusks (much reduced or absent in Aplacophora and in some pelecypods), often modified to form specialized structures, such as the arms of cephalopods or the swimming paddles of pteropods.

FOSSA: A trenchlike depression; in some gastropods (e.g., *Nassarius*), a spiral groove separating the base of the shell from the anterior canal.

FOSSETTE: A pelecypod resilifer that is pitlike in form. *See also* Chondrophore.

FURROW: A pronounced groove in a shell.

FUSED: Merged into a single structure or surface.

FUSIFORM: Spindle-shaped, with a long spire and anterior canal.

GAPE: A chink or a wider opening between closed pelecypod valves.

GAPING: Incapable of closing completely.

GILL RHACHIS: *See* Rhachis.

GIRDLE: A flexible, leathery, muscular integument surrounding the valves of a chiton, often ornamented with scales, spicules, or hairy processes.

GLOBOSE: Roughly spherical in outline.

GRANULAR: Bearing granules as surface sculpture.

GRANULE: A pustular surface structure.

GROOVE: An elongate and fairly uniform depression in the shell or soft parts of a mollusk.

GROWTH LINE: A line on the shell surface that marks the position of the shell's margin at a previous stage of growth, not sufficiently raised above the shell surface to be regarded as true sculpture.

GUTTERED: Marked with a wide, shallow groove or grooves.

HEAD: In general, the area of a molluscan body that bears the sense organs and the mouth (or proboscis). Pelecypods have no head as such, though the mantle edge may have light-sensitive organs.

HEAD VALVE: In chitons, the valve at the anterior end of the body.

HEIGHT: The greatest vertical dimension: in gastropods, the dimension parallel to the axis of coiling; in pelecypods, the measure along a line passing through the beak and perpendicular to a line bisecting the adductor muscles.

HELICAL: Spirally coiled.

HETERODONT: In pelecypods, having distinctly differentiated cardinal and lateral teeth on the hinge.

HETEROSTROPHIC: Said of coiling such that the initial whorls seem to turn opposite to the direction of coiling of later whorls.

HINGE: Collective term for structures of the dorsal region of pelecypod shells that function in opening and closing the valves.

HINGE LINE: *See* Hinge plate.

HINGE PLATE: In pelecypods, the infolded dorsal margin of a valve, carrying the hinge teeth.

HINGE TEETH: Shelly structures (usually in series) along or under the dorsal margin of a pelecypod shell, fitting into sockets in the opposite valve and serving to assure accurate closure.

IMMERSED: Condition of apical whorls in a gastropod shell when sunk within later ones and partially or entirely concealed by them.

IMPRESSED: Pushed down, either as a line or an area.

INCISED: Sculptured with one or more sharply cut grooves.

INCREMENTAL LINES: Faint concentric growth lines.

INCRUSTATION: An irregular deposit on the shell surface.

INEQUILATERAL: In pelecypods, having the anterior and posterior sections of the valves dissimilar in shape and size.

INEQUIVALVE: In pelecypods, having the two valves dissimilar in shape and size.

INFLATED: Swollen.

INITIAL WHORL: In gastropods, the first visible whorl at the apex.

INSERTION PLATE: In chitons, flangelike structures at the sides of each valve, covered by the girdle in the living animal.

INSERTION TEETH: In chitons, flattened serrations on the insertion plates, projecting into the girdle.

INTEGUMENT: An outer covering layer of molluscan soft parts; epidermis; cuticle.

INTERNAL LIGAMENT: *See* Ligament; Resilium.

INTERSPACES: Channels between sculptural ribs.

INVOLUTE: In gastropods (e.g., *Cypraea*), coiling so that later whorls (or just the final whorl) envelop earlier whorls, such that the height of the aperture is the greatest vertical dimension.

JUGUM: In chitons, the raised area in the central top surface of a valve.

KEEL: A prominent, sharply raised rib, generally marking an abrupt change of slope in the shell outline; same as carina.

LAMELLA (pl. LAMELLAE) or LAMINA: A thin plate or scale; in some pelecypods, one of the small, raised, regularly arranged processes (equivalent to teeth) on the hinge plates of the valves.

LARVAL SHELL: Shell of a molluscan larva before it undergoes metamorphosis, usually set off by a change of sculpture.

LATERAL AREA: In chitons, the dorsal surface at either side of a valve.

LATERAL TEETH: In heterodont pelecypods, hinge teeth posterior to or anterior to the cardinal teeth (*q.v.*); in gastropods, radular elements that lie between the central and marginal elements in each row.

LENGTH: In pelecypods, the greatest horizontal dimension parallel to a line bisecting the adductor scars; in gastropods and scaphopods, the same dimension as height; in chitons, the greatest dimension parallel to the axis of symmetry; in cephalopods, variously defined, but generally taken as the greatest dimension of the animal when the arms are stretched out parallel to their full extent.

LENTICULAR: Lens-shaped; i.e., flattened and generally circular.

LIGAMENT: In pelecypods, a horny elastic structure or structures joining the two valves dorsally, mostly posterior to the beaks, and acting as a spring that causes the valves to open when adductor muscles relax, the external ligament being under tension, the internal (sometimes called "cartilage" or "resilium") being under compression.

LIP: In gastropods, the margin of the aperture: the inner lip (labium) extends from the base of the columella to the suture and is divided into a columellar lip and a parietal lip; the outer lip (labrum) is that part of the lip furthest from the axis of coiling.

LIRA (pl. LIRAE): A threadlike sculptural element.

LIRATE: Having a threadlike sculpture.

LITHODESMA: A small calcareous plate reinforcing an internal ligament in some pelecypods.

LONGITUDINAL: Along the length (e.g., of an apparently uncoiled gastropod shell). Not commonly used in the literature because of inconsistencies. *See also* Transverse.

LUNULE: In pelecypods, a heart-shaped area on the dorsal margin just anterior to the beaks, set off by a differing sculpture.

MALLEATION: Shell sculpture resembling a hammered metal surface.

MANDIBLE: One element of a cephalopod beak or jaw.

MANTLE: The fleshy outer layer of a molluscan body that secretes the shell and periostracum, also sometimes forming the tubular folds (the siphons) that circulate water for respiration and feeding.

MANTLE SCAR: A broad area of attachment of the mantle edge in pelecypods (e.g., *Sportella*), too wide to be called a pallial line.

MARGIN: The edge, particularly of a shell surface.

MARGINAL TEETH OR MARGINALS: The outermost series of teeth in one transverse row of the molluscan radula.

MEDIAL: Central.

MEDIAN OR MIDLINE: A central line or axis.

MUCRO: In chitons, the raised apex on the dorsal surface of the tail valve.

MULTISPIRAL: With many turns; especially as applied to gastropod opercula.

MUSCLE SCAR: An impression on a shell surface marking the attachment of muscles; in limpets, the scar left by the margin of the mantle and the foot retractor muscles. *See also* Accessory scar; Adductor scar.

MYOPHORE: A shelly structure that projects into the cavity of a pelecypod valve (e.g., in the Pholadidae) and provides attachment for foot retractor muscles. *See also* Apophysis.

NACREOUS: Pearly; having an iridescent luster and a special composition of calcium carbonate and organic material.

NEST: A dwelling chamber constructed by certain pelecypods, of agglutinated sand and rubble.

NESTLING: Fixed in a crevice or crack.

NODE: A small knob or knot.

NODOSE: Sculptured with nodes.

NUCLEAR TIP: The apex of a gastropod shell.

NUCLEAR WHORL: One of the whorls forming the gastropod nucleus.

NUCLEUS: In gastropods, the initial whorls of the shell (protoconch) or the first-formed portion of an operculum; in pelecypods, the prodissoconch.

NYMPH: In pelecypods, a thickened projection along the hinge margin that supports an external ligament or reinforces the normal hinge structure, as in *Gari*.

OBCONIC: Approximately cone-shaped.

OBLIQUE: Sculptured other than axially, radially, or concentrically; i.e., other than parallel or perpendicular to a growing edge.

OBOVATE: Reversed ovate; having the greatest shell width above the aperture and toward the apex.

OBSOLETE: Said of a structure or sculpture that tends to disappear or remain undeveloped.

OPERCULUM (pl. OPERCULA): A corneous or calcareous structure attached to the foot of a gastropod, serving to close the aperture when the animal withdraws into its shell.

OPISTHOGYRATE: In pelecypods, having the beaks pointed backward, or posteriorly; opposite of prosogyrate.

ORBICULAR: Circular.

ORIFICE: An opening.

ORTHOGYRATE: In pelecypods, having the beaks pointed toward each other.

OVATE or OVOID: Having an oval shape.

PALLETS: A pair of calcareous structures at the siphonal end of some wood-boring pelecypods, used to close the opening of the burrow.

PALLIAL LINE: In pelecypods, a linear impression connecting the adductor muscle scars and marking the attachment line of the muscles of the mantle edge to the shell. *See also* Mantle scar.

PALLIAL MUSCLE: A muscle associated with the mantle of a pelecypod.

PALLIAL SINUS: In pelecypods, an embayment of the pallial line, marking the space occupied by the retractor muscles for the siphons, hence always on the posterior part of the shell.

PARASITIC: Living at the expense of a host species. *See also* Commensal.
PARIETAL: In gastropods, referring to the basal area of a helically coiled shell, just inside and outside the aperture and above the columella.
PATELLATE: Saucer-shaped or limpet-shaped.
PAUCISPIRAL: Having few whorls. *See also* Multispiral.
PEBBLED: Sculptured with small, irregular bumps.
PEDAL: Relating to the molluscan foot.
PEDAL GAPE: In pelecypods, a space between the valves through which the foot protrudes.
PEDAL GLAND: A gland on the foot, especially in gastropods, secreting, for example, a lubricating substance to aid locomotion.
PELAGIC: Living in the open sea, especially near the surface.
PELLUCID: Transparent or translucent.
PEN: The internal stiffening rod of some cephalopods.
PERFORATION: A small hole.
PERIOSTRACUM: The outermost layer of a molluscan shell, composed of conchiolin in various textures; erroneously called "epidermis."
PERIPHERY: In gastropods, the part of any shell or any individual whorl that is farthest from the shell's axis of coiling.
PERISTOME: In gastropods, the margin of the aperture, said to be entire or complete when it is not interrupted, as by the parietal area of the body whorl.
PERSISTENT: Lasting throughout the development of a shell.
PILLAR: Same as columella; also the part of the body whorl to the left of the columellar lip.
PLAIT: *See* Plication.
PLANISPIRAL: Coiled in a single plane.
PLANORBOID: Having a flattened planispiral coiling, the name deriving from the shells of the freshwater gastropod family Planorbidae.
PLICATE: Folded or twisted.
PLICATION: A raised ridge, fold, or plait, especially on the columella of a gastropod shell.
PLUG: *See* Apical plug.
PORCELANEOUS: Having a translucent, porcelain-like appearance.
POSTAPICAL WHORL: In gastropods, the whorl immediately adjoining the apical whorl.
POSTERIOR: In gastropods, the direction opposite to that in which the head points when the animal is active; in pelecypods, the direction in which the ligament, siphons, and pallial sinus lie; in chitons and cephalopods, the end opposite that on which the mouth and radula occur; the small end of scaphopods. *See also* Anterior; Dorsal; Ventral.
POSTNUCLEAR WHORL: In gastropods, any whorl other than those immediately associated with the nucleus.
PROCESS: Part of a shell or organism that projects outward from the main mass or body.

PRODISSOCONCH: In pelecypods, the shell secreted by the larva, preserved at the beak of some adult shells (see drawing of *Bernardina*). *See also* Nucleus.

PRODUCED: Drawn out; elongate.

PROSOGYRATE: Turned forward; in pelecypod shells, having the beaks directed anteriorly; opposite of opisthogyrate.

PROTOCONCH: In gastropods, the apical or nuclear whorls of a shell, generally clearly demarcated from the later whorls. *See also* Nucleus.

PUNCTATE: Having minute pits.

PUSTULE: A unit of knobby sculpture, generally smaller than a tubercle.

QUADRATE: Squarish or rectangular in outline.

RADIAL: Radiating in relatively unbroken lines from the beaks of pelecypods, the apex of gastropods, or the mucro of chitons, as contrasted with concentric.

RADULA (pl. RADULAS or RADULAE): A rasplike structure in the mouth of all molluscan groups except pelecypods, composed of chitinous material and arranged as a long, coiled band set with serial or transverse rows of teeth, each transverse row generally consisting of central, lateral, and marginal teeth.

RECURVED: Hooked or bent; said of the siphonal canal in some gastropods, which may be turned so far backward that its end is not visible when the shell is held in normal orientation.

RESILIFER: In pelecypods, a structure that supports the internal ligament. *See also* Chondrophore, Fossette.

RESILIUM: In pelecypods, the internal ligamental portion of the hinge. *See also* Cartilage; Ligament.

RESTING STAGE: A point at which the molluscan shell is not being enlarged to accommodate growth, often marked by a thickening or ridge left on the shell surface.

RETICULATE: *See* Cancellate.

RHACHIS: A stemlike central gill axis supporting other elements.

RHIPIDOGLOSSATE: Having a radular dentition in which the marginal teeth are numerous, resembling the ribs of a fan.

RIB: An elongated structural element: in gastropods, spiral (following the direction of coiling), axial (parallel to the axis of coiling), or oblique; in pelecypods, radial or concentric. *See also* Costa.

RIBLET: A narrow rib.

RIMATE: Fissured; with a chink.

ROSTRATE: Drawn out into a beaklike process.

SCALE: A sharp, raised element in shell sculpture; ornamentation on the girdle of chitons.

SCAR: A marking on the interior of a shell that indicates the attachment point of a muscle.

SEPTUM: A transverse plate of shelly material. *See also* Deck; Shelf.

SERRATE: Notched or sawlike.

SESSILE: Attached to some substrate; not free to move about.

SETA (pl. SETAE): A bristle-like structure.

SHELF: A plate of shelly material, e.g., the deck (*q.v.*) in *Crepidula*. *See also* Septum.

SHELLY: Composed of calcium carbonate rather than conchiolin; i.e., with a calcareous rather than a horny texture.

SHOULDER: In gastropods (e.g., *Acanthina*), an angulation of the whorl at the periphery, forming the outer edge of a sutural ramp or shelf.

SINISTRAL: Coiling in a counterclockwise direction. *See also* Dextral.

SINUOUS or SINUATE: Having the margin wavy.

SINUS: A bend or embayment in growth lines, in the outer lip of a gastropod, or in the attachment scar of the mantle in pelecypods.

SIPHON: A tubelike structure (or one of two or more such structures), formed by folds or fusions of the mantle edge, that directs water currents through the molluscan body.

SIPHONAL FASCIOLE: In gastropods, a spiral, roughened band near the anterior end of the columella, marking successive positions of the siphonal notch.

SLIT BAND: In gastropods, a spiral band of crescentic growth lines ascending the whorls of the shell and marking the successive positions of a slit in the outer lip of the aperture. Also called (in other works) the selenizone.

SLOPE: One face of an angled bivalve shell, whether anterior, posterior, or central.

SOCKET: In pelecypods, a recess for the reception of a hinge tooth or a chondrophore from the opposite valve.

SPICULE: A hard needlelike body in the integument (especially in chitons), serving as stiffening.

SPINOSE: Having spines or thornlike protuberances.

SPIRE: In gastropods, the visible part of all whorls except the final, or body, whorl.

STRIA (pl. STRIAE): A line or ring indicating a growth stage.

STRIATE: Marked with a sculpture of fine scratches, grooves, or lines.

SUBDUED: Weak, not evident.

SUBSTRATE: The sea floor or other base for attachment, crawling, or burrowing by a mollusk.

SULCUS: A deep slit, fissure, or furrow.

SUNKEN: Indented or depressed. *See also* Impressed.

SUTURE: In gastropods, the spiral line that marks the junction of the whorls; in chitons, the junction between girdle and valves.

TABULATE: Strongly shouldered; said of a formation near the suture of gastropod whorls (e.g., in *Busycon*) in which the later whorl meets the previous whorl at approximately right angles, forming a flattened, horizontal ramp often bounded by a low carina.

TAIL VALVE: In chitons, the most posterior of the eight valves, differing in shape from the others.

TAXODONT: In pelecypods, a hinge dentition composed of alternating teeth and sockets, mostly similar in form, in a series of varying length.

TEETH: Interlocking projections from the dorsal margins of pelecypod valves;

projecting nodes in the aperture of gastropods; flattened serrations of the margins of chiton valves; the elements of the radula. The hinge teeth or raised lamellae on the hinge plates of pelecypods are arranged in regular patterns and are important in identifications.

TERMINAL: At the extreme end, especially with reference to a cylindrical or long-ovate pelecypod shell.

THREAD: A thin, elongated element in surface sculpture; lira; also, in pelecypods, one of the structures secreted by a byssal gland for attachment to the substrate.

TRANSVERSE: At right angles to longitudinal (used in a general sense but also with reference to the serial rows of teeth on some radulae).

TRAPEZOIDAL: Having a quadrate outline with one longitudinal dimension greater than the other.

TRIGONAL: Triangular in outline.

TROCHOID: Top-shaped (as in *Calliostoma*).

TRUNCATE: Sharply or squarely cut off.

TUBERCLE: A knob or lump.

TURBINATE: Turban-shaped; trochoid, but with the whorls more strongly inflated (e.g., *Tegula*).

TURRETED: In gastropods, tower-shaped, with a long spire and somewhat shouldered whorls.

TURRIFORM: With a many-whorled, slender spire.

ULTRADEXTRAL: Having seemingly sinistral shell coiling; having the soft parts dextral, but with shell whorls being added above instead of below the periphery; hyperstrophic.

UMBILICATE: Having an umbilicus.

UMBILICUS: An open axis of coiling in a spiral gastropod shell.

UMBO (pl. UMBONES): The upper (or earliest) part of a pelecypod valve, as seen from the outside, the youngest, terminal portion of which is the beak, best seen in an interior view of the valve.

UNCINUS (pl. UNCINI): In gastropods, a marginal or lateral tooth on the radula, more or less hooklike in shape.

VALVE: In pelecypods, one of the two portions into which the shell is divided, the two valves usually joined by a hinge.

VARIX (pl. VARICES): An elevated axial structure in certain gastropods, more prominent than ribs and generally more widely spaced, marking periodic resting stages in the animal's growth, during each of which a thickened outer lip developed.

VENTRAL: Pertaining to, in gastropods, that part of the shell that is lowermost when the animal is extended, i.e. the apertural face; in pelecypods, the margin farthest away from the umbo; in chitons, the foot surface. *See also* Dorsal; Anterior; Posterior.

VERMICULATION: A surface sculpture of irregular wavy lines or grooves.

VOLUTION: Any complete turn in a spirally wound gastropod shell.

WHORL: In gastropod spires, an exposed portion of a volution between successive sutures.

WIDTH: The greatest dimension at right angles to length or height.

WING (or WINGLET): A more or less elongate triangular extension of the hinge area in some pelecypod groups, or of the apertural margin in some gastropods.

B. Terms Used in Scientific Nomenclature

AUCTT.: *Auctorum* ("of authors"): an abbreviation indicating that the usage is not now regarded as appropriate for the cited taxon, although the name itself may be valid (i.e., the term indicates the misuse of a name).

EMEND.: Marks an emendation or deliberate alteration in the spelling of a name, usually an unjustified alteration.

EX: "From," as "*ex* Doe, MS."

FIDE: On the authority of (a published authority).

HOLOTYPE: The specimen upon which a name for a species is based.

HOMONYM: The later of two identical names that were applied to two different genera or species.

ICZN: International Commission on Zoological Nomenclature, an advisory group established by the International Zoological Congress to formulate a nomenclatural code and to render opinions in response to petitions by individual systematists.

IN LITT.: In correspondence.

MS (pl. MSS): Manuscript, unpublished work.

NOM. CONSERV.: *Nomen conservandum*: a name officially preserved by the ICZN in the interests of clarity or stability.

PREOCCUPIED: Invalid because some author has previously used the same name for a different animal.

S.L.: *Sensu lato*: in the broad sense.

S.S.: *Sensu stricto*: in the strict sense.

SYNONYM: Customarily, the later of two different names that have been given to a single taxon; technically, both are synonyms, the earlier being the senior and the later the junior synonym.

TAXON (pl. TAXA): Any unit in classification, irrespective of rank; for example, a generic taxon is any unit treated as a rank between family and species, with no implication of whether it has generic or subgeneric status.

TESTE: According to (verbal or unpublished testimony). *See also* Fide.

UNJUSTIFIED EMENDATION: Any deliberate alteration of an original spelling of a name; only a few causes for emendation are justified under the ICZN Code.

BIBLIOGRAPHY

Desirable as a complete bibliography on Western North American Mollusca would be, such a project is beyond the scope of the present work. Some parts of the task have been accomplished, for Dall in 1909 published a fairly comprehensive list of works up to that date; Keen (1937, 1956) carried the compilation further. The present bibliography stresses the publications of the period since then, with emphasis on the books and papers to which the student may turn for identification of species once the probable genus has been determined by use of the keys. In general, the papers cited here are those having bibliographies that will lead the student to earlier literature.

A topical summary may be useful here to group the titles into pertinent categories:

General works. W. Arnold, 1966; Hanna, 1939, 1966; ICZN, 1961; Keen, 1937, 1956; Moore et al., 1960, 1969-71; Taylor & Sohl, 1962; Vokes, 1967.

Identification manuals (covering both pelecypods and gastropods). Abbott, 1954, 1968; Keen, 1971; Keen & Frizzell, 1939; Keen & Pearson, 1952; Keep, 1935; MacGinitie, 1959; McLean, 1969; Morris, 1966; Oldroyd, 1925-27; Palmer, 1958; Rice, 1972.

Identification aids. Gastropoda (other than opisthobranchs): Coan & Roth, 1966; Cox, 1962; McLean, 1967*a-b*, 1971; Olsson, 1956; Rokop, 1972; Roth & Coan, 1968; Strauch, 1972; Tesch, 1949; Tikasingh & Pratt, 1961. *Gastropoda (opisthobranchs)*: Beeman, 1968; Bertsch et al., 1972; Gosliner & Williams, 1970; Lance, 1961, 1966; MacFarland, 1966; McGowan, 1968; Marcus, 1961, 1967; Odhner, 1963; Roller, 1970; Roller & Long, 1969; Sphon, 1972; Steinberg, 1963*a-c*; Thompson, 1971; Van der Spoel, 1967. *Pelecypoda (Bivalvia)*: Bernard, 1969, 1972; Coan, 1971, 1973*a-c*; Fitch, 1953; Grau, 1959; Harry, 1969; Knudsen, 1970; MacNeil, 1965, 1967; Olsson, 1961; Rost, 1955; Soot-Ryen, 1955; Strauch, 1972; Swan & Finucane, 1952; Turner, 1954-55, 1966. *Scaphopoda*: Emerson, 1962. *Polyplacophora*: Burghardt & Burghardt, 1969. *Aplacophora*: Heath, 1911; Schwabl, 1963. *Cephalopoda*: Berry, 1912; Clarke, 1966; Pearcy, 1965; Roper, 1969; Roper et al., 1969; Voss, 1970; Young, 1972.

Distribution lists for Western North America. Bernard, 1970; Burch, 1944-46; Coan, 1964; Dall, 1921; Gosliner & Williams, 1970; Lance, 1961; MacDonald, 1969; MacGinitie, 1959; McLean, 1969; Ricketts & Calvin, 1968; Roller & Long, 1970; Smith & Gordon, 1948; Sphon, 1972; Steinberg, 1963*a-c*.

Pleistocene and late Pliocene mollusks. Addicott, 1973; Arnold, 1903; Grant & Gale, 1931; Hertlein & Grant, 1972.

Many of the gastropods of the West Coast are of such small size that they must be studied with a microscope. Their identification is a special field. To facilitate such identifications, the following list of papers has been adapted from one first published by A. M. Strong in 1934. Most of the papers are by Paul Bartsch; unless otherwise noted, the condensed citations here are to the Proceedings of the U.S. National Museum, volume number, publication number, and (in parentheses) date. Titles given in conventional fashion are in the Bibliography proper. Arrangement is alphabetical by genera and family groups covered:

Alaba: 39/1781 (1910). *Alabina*: 39/1790 (1911). *Alvinia* (as *Alvania*): 41/1863 (1911). *Amphithalamus*: 41/1854 (1911). *Assiminea* (as *Syncera*): 58/2331 (1920). *Balcis*: see *Eulima*. *Barleeia*: 58/2331 (1920). *Bittium*: 40/1826 (1911). *Cerithiopsis*: 40/1823 (1911). *Cingula* (including *Nodulus*): 41/1858 (1911); 41/1871 (1912). *Diastoma*: 39/1802 (1911). *Eulima* and *Balcis* (as *Melanella*): 53/2207 (1917). Marginellidae: Coan & Roth, 1966; Roth & Coan, 1968. *Odostomia*: Dall & Bartsch, 1909. *Rissoella*: 58/2331 (1920). *Rissoina*: 49/2049 (1915). *Tricolia* (as *Phasianella*): Strong, 1928. *Turbonilla*: Dall & Bartsch, 1909. Vitrinellidae: Pilsbry & Olsson, 1945, 1952. *Volvulella*: Harry, 1967.

The Bibliography proper follows. Several of the larger works listed are out of print (†) but are still obtainable in public or institutional libraries.

Abbott, R. Tucker. 1954. American seashells. New York, xiv + 541 pp., 310 figs., 40 pls. (Rev. ed. in press.)

——— 1968. A guide to field identification: Seashells of North America. Racine, 280 pp., numerous col. figs.

Addicott, W. O. 1965. Some Western American Cenozoic gastropods of the genus *Nassarius*. U.S. Geol. Survey Prof. Paper 503-B, 24 pp., 3 pls.

——— 1973. Neogene marine mollusks of the Pacific Coast of North America: An annotated bibliography, 1797-1969. U.S. Geol. Surv. Bull. 1632. 201 pp.

Arnold, Ralph. 1903. The paleontology and stratigraphy of the marine Pliocene and Pleistocene of San Pedro, California. Mem. Calif. Acad. Sci., vol. 3, 420 pp., 37 pls.†

Arnold, Winifred. 1966. A glossary of a thousand-and-one terms used in conchology. Veliger, vol. 7 (suppl.), 50 pp., 155 figs.

Beeman, R. D. 1968. The order Anaspidea. Veliger, vol. 3 (suppl.), pt. 2, pp. 87-102, 12 figs., 1 pl.

Bernard, Frank R. 1969. Preliminary diagnoses of new septibranch species from the eastern Pacific (Bivalvia: Anomalodesmata). J. Fish. Res. Bd. Canad., vol. 26, no. 8, pp. 2230-34, 1 pl.

——— 1970. A distributional checklist of the marine molluscs of British Columbia, based on faunistic surveys since 1950. Syesis, vol. 3, 75-94, 1 pl.

——— 1972. The genus *Thyasira* in western Canada (Bivalvia: Lucinacea). Malacologia, vol. 11, no. 2, pp. 365-89, 17 figs.

Berry, S. S. 1912. A review of the cephalopods of western North America. Bull. U.S. Bur. Fish., vol. 30, doc. 761, pp. 267-336, pls. 32-56, 18 text figs.

Bertsch, Hans, T. Gosliner, R. Wharton, and G. Williams. 1972. Natural history and occurrence of opisthobranch gastropods from the open coast of San Mateo County, California. Veliger, vol. 14, no. 3, pp. 302-14.

Burch, J. Q., ed. 1944-46. Distributional list of the West American marine Mollusca from San Diego, California, to the Polar Sea. Extracts from the Minutes of the Conchological Club of Southern California, pt. I (Pelecypoda), nos. 33-45 (Mar. 1944-Feb. 1945); pt. II, vols. I & II (Gastropoda), nos. 46-63 (Mar. 1945-Sept. 1946).†

Burghardt, Glenn, and Laura Burghardt. 1969. A collector's guide to West Coast chitons. San Francisco Aquarium Soc., Spec. Publ., no. 4, 45 pp., 4 col. pls., 7 text figs.

Clarke, M. R. 1966. A review of the systematics and ecology of oceanic squids. Advances in Marine Biology, vol. 4, pp. 91-300, 59 figs.

Coan, E. V. 1964. The Mollusca of the Santa Barbara County area, pt. I: Pelecypoda and Scaphopoda. Veliger, vol. 7, no. 1, pp. 29-33 (Oct.).

——— 1971. The Northwest American Tellinidae. Veliger, vol. 14 (suppl.), 63 pp., 12 pls., 30 figs.

——— 1973*a*. The Northwest American Semelidae. Veliger, vol. 15, no. 4, pp. 314-329, 2 pls. (Apr. 1).

——— 1973*b*. The Northwest American Psammobiidae. Veliger, vol. 16, no. 1, pp. 40-57, 4 pls. (July 1).

——— 1973*c*. The Northwest American Donacidae. Veliger, vol. 16, no. 2, pp. 130-39, 1 pl., 2 text figs. (Oct. 1).

Coan, E. V., and Barry Roth. 1966. The West American Marginellidae. Veliger, vol. 8, no. 4, pp. 279-99, pls. 48-51, 5 figs.

Cox, K. W. 1962. California abalones, family Haliotidae. Calif. Dept. Fish & Game, Fish. Bull. no. 118, 133 pp., 61 figs.

Dall, W. H. 1896. Pelecypoda, in K. A. von Zittel, Text-book of paleontology (tr. & ed. by C. R. Eastman), rev. English. ed., London, 1900, vol. 1, pt. 1, pp. 346-429, figs. 589-781, 1 tbl. (Separately issued in advance of the work.)†

——— 1921. Summary of the marine shellbearing mollusks of the northwest coast of America. Bull. U.S. Nat. Mus., no. 112, 217 pp., 22 pls.

Dall, W. H., and Paul Bartsch. 1909. A monograph of West American pyramidellid mollusks. Bull. U.S. Nat. Mus., no. 68, 258 pp., 30 pls.

Emerson, W. K. 1962. A classification of the scaphopod mollusks. Jour. Paleontol., vol. 36, no. 3, pp. 461-82, pls. 76-80, 2 figs.

Fitch, J. E. 1953. Common marine bivalves of California. Calif. Dept. Fish & Game, Fish. Bull., no. 90, 102 pp., 1 pl., 63 figs.

Gainey, L. F. 1972. The use of the foot and the captacula in the feeding of *Dentalium*. Veliger, vol. 15, no. 1, pp. 29-34, 5 figs.

Gosliner, T. M., and G. C. Williams. 1970. The opisthobranch mollusks of Marin County, California. Veliger, vol. 13, no. 2, pp. 175-80, 1 map.

Grant, U.S., IV, and H. R. Gale. 1931. Catalogue of the marine Pliocene and Pleistocene Mollusca of California. Mem. San Diego Soc. Nat. Hist., vol. 1, 1036 pp., 32 pls., 15 figs.

Grau, Gilbert. 1959. Pectinidae of the eastern Pacific. Univ. So. Calif. Publ. (Allan Hancock Pacific Expeditions), no. 23, 308 pp., 57 pls.

Hanna, G. 1939. Exotic Mollusca in California. Bull. Calif. Dept. Agric., vol. 28, no. 5, pp. 298-321, 4 pls., 2 figs.

——— 1966. Introduced mollusks of western North America. Occ. Papers Calif. Acad. Sci., no. 48, 108 pp., 4 pls., 85 figs.

Harry, H. W. 1967. A review of the living tectibranch snails of the genus *Volvulella*. Veliger, vol. 10, no. 2, pp. 133-47, 21 figs.

——— 1969. A review of the living leptonacean bivalves of the genus *Aligena*. Veliger, vol. 11, no. 3, pp. 164-81, 40 figs.

Heath, Harold. 1911. The solenogastres. Mem. Harvard Mus. Comp. Zool., vol. 45, no. 1, pp. 1-182, 40 pls.

Hertlein, L. G. 1959. Notes on California oysters. Veliger, vol. 2, no. 1, pp. 5-10, 1 pl.

——— 1961. A new species of *Siliqua* (Pelecypoda) from western North America. Bull. So. Calif. Acad. Sci., vol. 60, no. 1, pp. 12-19, 2 pls.

Hertlein, L. G., and U. S. Grant, IV. 1972. The geology and paleontology of the marine Pliocene of San Diego, California (Paleontology: Pelecypoda). San Diego Soc. Nat. Hist., Mem. 2, pt. 2B, pp. 135-511, pls. 27-57, figs. 7-13.

International Code of Zoological Nomenclature adopted by the XV International Congress of Zoology. 1961. London. International Trust for Zoological Nomenclature, xvii + 176 pp. Reprinted with minor additions 1964. (In English and French. Obtainable for $3.00 from the Publications Office of the International Trust, 14 Belgrave Square, London, SWIX-8PS, England.)

Keen, A. M. 1937. An abridged checklist and bibliography of West North American marine Mollusca. Stanford, Calif.: Stanford Univ. Press. 87 pp.†

——— 1956. Supplement to the above (Papers published 1937-56). Mimeo., available from the author. 13 pp.

——— 1971. Sea shells of Tropical West America (2d ed.). Stanford, Calif.: Stanford Univ. Press. xiv + 1064 pp., 22 col. pls., about 4,000 figs., 6 maps.

Keen, A. M., and D. L. Frizzell. 1939. Illustrated key to West North American pelecypod genera. Stanford, Calif.: Stanford Univ. Press. 28 pp., 147 figs. (rev. ed. 1953, 32 pp., 149 figs.).†

Keen, A. M., and J. C. Pearson. 1952. Illustrated key to West North American gastropod genera. Stanford, Calif.: Stanford Univ. Press. 39 pp., 190 figs.†

Keep, Josiah. 1935. West Coast shells (rev. by J. L. Baily, Jr.). Stanford, Calif.: Stanford Univ. Press. xi + 350 pp., 334 figs.†

Knudsen, Jørgen. 1970. The systematics and biology of abyssal and hadal Bivalvia. Galathea Rept., vol. 11, 241 pp., 20 pls., 132 figs.

Kozloff, Eugene. 1973. Seashore life of Puget Sound, the Strait of Georgia, and the San Juan Archipelago. Seattle: University of Washington Press. viii + 282 pp., 28 col. pls., 223 photos & ln. drwngs.

Lance, J. R. 1961. A distributional list of southern Californian opisthobranchs. Veliger, vol. 4, no. 2, pp. 64-69.

——— 1966. New distributional records of some northeastern Pacific Opisthobranchiata. Veliger, vol. 9, no. 1, pp. 69-81, 12 figs.

MacDonald, K. B. 1969. Molluscan faunas of Pacific coast salt marshes and tidal creeks. Veliger, vol. 11, no. 4, pp. 399-407, 1 map.

MacFarland, F. M. 1966. Studies of opisthobranchiate mollusks of the Pacific Coast of North America. Mem. Calif. Acad. Sci., vol. 6, 546 pp., 71 pls.

MacGinitie, Nettie. 1959. Marine Mollusca of Point Barrow, Alaska. Proc. U.S. Nat. Mus., vol. 109, no. 3412, pp. 59-208, 27 pls.

MacNeil, F. S. 1965. Evolution and distribution of the genus *Mya* and Tertiary migrations of Mollusca. U.S. Geol. Survey Prof. Paper 483-G, iv + 51 pp.

——— 1967. Cenozoic pectinids of Alaska, Iceland, and other northern regions. U.S. Geol. Survey Prof. Paper 553, 57 pp., 25 pls.

McGowan, J. A. 1968. Thecosomata and Gymnosomata. Veliger, vol. 3 (suppl.), pt. 2, pp. 103-129, pls. 12-20.

McLean, J. H. 1967*a*. West American species of *Lucapinella*. Veliger, vol. 9, no. 3, pp. 349-52, pl. 49, 3 figs.

——— 1967*b*. West American Scissurellidae. Veliger, vol. 9, no. 4, pp. 404-10, pl. 56.

——— 1969. Marine shells of southern California. Los Angeles County Mus. Nat. Hist., Sci. Ser. no. 24, Zool. no. 11, 108 pp., 54 figs.

——— 1971. A revised classification of the family Turridae. Veliger, vol. 14, no. 1, pp. 114-30, 4 pls.

Marcus, Ernst. 1961. Opisthobranch mollusks from California. Veliger, vol. 3 (suppl.), pt. 1, 85 pp., 10 pls.

Marcus, Ernst, and Eveline Marcus. 1967. American opisthobranch mollusks. Stud. Tropical Oceanography, no. 6, viii + 256 pp., 1 pl., 240 figs.

Moore, R. C., ed. 1960. Treatise on invertebrate paleontology. Part I: Mollusca 1, 351 pp., 216 figs. (Mollusca, General features; Scaphopoda; Amphineura; Monoplacophora; Archaeogastropoda). Geol. Soc. America and Univ. Kansas Press.

——— 1969-71. Treatise on invertebrate paleontology. Part N: Mollusca 6, (Bivalvia), 3 vols. Vols. 1-2, 952 pp.; vol. 3, pp. 953-1224 (Ostreidae). Geol. Soc. America and Univ. Kansas Press.

Morris, Percy. 1966. A field guide to shells of the Pacific Coast and Hawaii. 2d ed. Boston. xxxiii + 297 pp., 72 pls. (8 col.).

Odhner, N. H. 1963. On the taxonomy of the family Tritoniidae (Mollusca: Opisthobranchia). Veliger, vol. 6, no. 1, pp. 48-52.

Oldroyd, I. S. 1925-27. The marine shells of the West Coast of North America (Stanford Univ. Pub. Geol.). Vol. 1, Pelecypoda (dated 1924, publ. 1925), 248 pp., 57 pls. Vol. 2, Gastropoda, Scaphopoda, and Amphineura: pt. 1, 298 pp., pls. 1-29; pt. 2, 304 pp., pls. 30-72; pt. 3, 340 pp., pls. 73-108.†

Olsson, A. A. 1956. Studies on the genus *Olivella*. Proc. Acad. Nat. Sci. Philadelphia, vol. 108, pp. 155-225, pls. 8-16.

——— 1961. Mollusks of the Tropical Eastern Pacific: Panamic-Pacific Pelecypoda. Ithaca, N.Y.: Paleontological Research Institution, 574 pp., 86 pls. (Includes illustrations and discussions of some Californian forms.)

Palmer, K. van W. 1958. Type specimens of marine Mollusca described by P. P. Carpenter from the West Coast (San Diego to British Columbia). Mem. Geol. Soc. Amer., no. 76, 376 pp., 35 pls.

Pearcy, W. G. 1965. Species composition and distribution of pelagic cephalopods from the Pacific Ocean off Oregon. Pac. Science, vol. 19, no. 2, pp. 261-66, 3 figs.

Pilsbry, H. A., and A. A. Olsson. 1945-52. Vitrinellidae and similar gastropods of the Panamic Province. Pts. I-II. I, Acad. Nat. Sci. Philadelphia, vol. 97, 249-78, pls. 22-30 (1945); II, *ibid.*, vol. 104, 35-88, pls. 2-13 (1952).

Rice, Tom. 1972. Marine shells of the Pacific Northwest. Port Gamble, Wash.: Sea & Shore. 102 pp., 40 col. pls., 3 figs.

Ricketts, E. F., and Jack Calvin (fourth ed., revised by Joel W. Hedgpeth). 1968. Between Pacific Tides. Stanford, California: Stanford Univ. Press. xvi + 614 pp., 8 col. pls., 302 figs.

Rokop, F. J. 1972. Notes on abyssal gastropods of the eastern Pacific, with descriptions of three new species. Veliger, vol. 15, no. 1, pp. 15-19, 2 pls.

Roller, R. A. 1970. A list of recommended nomenclatural changes for MacFarland's "Studies of opisthobranchiate mollusks of the Pacific coast of North America." Veliger, vol. 12, no. 3, pp. 371-74.

Roller, R. A., and S. J. Long. 1969. An annotated list of the opisthobranchs of San Luis Obispo County, California. Veliger, vol. 11, no. 4, pp. 424-30.

Roper, C. F. 1969. Systematics and zoogeography of the worldwide bathypelagic squid *Bathyteuthis* (Cephalopoda: Oegopsida). Bull. U.S. Nat. Mus., no. 291, 210 pp., 12 pls., 74 figs.

Roper, C. F., R. E. Young, and G. L. Voss. 1969. An illustrated key to the families of the order Teuthoidea (Cephalopoda). Smithsonian Contrib. Zool., vol. 13, 32 pp., 16 pls., 2 figs.

Rost, Helen. 1955. A report on the family Arcidae (Pelecypoda). Univ. So. Calif. Publ. (Allan Hancock Pacific Expeditions), vol. 20, no. 2, pp. 177-249, pls. 11-16, figs. 79-95.

Roth, Barry, and E. V. Coan. 1968. Further observations on the West American Marginellidae, with the description of two new species. Veliger, vol. 11, no. 1, pp. 62-69, 1 pl., 2 figs., 1 map.

Schwabl, Mathilde. 1963. Solenogaster mollusks from Southern California. Pac. Science, vol. 17, no. 3, pp. 261-81, 28 figs. (July).

Smith, A. G., and Mackenzie Gordon. 1948. The marine mollusks and brachiopods of Monterey Bay, California, and vicinity. Proc. Calif. Acad. Sci., ser. 4, vol. 26, no. 8, pp. 147-245, 3 pls., 4 figs. (Dec. 15).

Smith, J. T. 1970. Taxonomy, distribution, and phylogeny of the cymatiid gastropods *Argobuccinum, Fusitriton, Mediargo*, and *Priene*. Bulls. Amer. Paleo., vol. 56, no. 254, pp. 445-573, pls. 39-49, 12 figs. (Apr. 2).

Soot-Ryen, Tron. 1955. A report on the family Mytilidae (Pelecypoda). Univ. So. Calif. Publ. (Allan Hancock Pacific Expeditions), vol. 20, no. 1, 175 pp., 10 pls., 78 figs. (Nov. 10).

Sphon, G. G. 1972. Some opisthobranchs (Mollusca: Gastropoda) from Oregon. Veliger, vol. 15, no. 2, pp. 153-57, 1 fig. (Oct.).

Steinberg, J. E. 1963*a*. Notes on the opisthobranchs of the West Coast of North America, II: The Order Cephalaspidea from San Diego to Vancouver Island. Veliger, vol. 5, no. 3, pp. 114-17 (Jan. 1).

——— 1963*b*. Notes, III: Further nomenclatural changes in the Order Nudibranchia. *Ibid.*, vol. 6, no. 2, pp. 63-67 (Oct. 1).

——— 1963*c*. Notes, IV: A distributional list of opisthobranchs from Point Conception to Vancouver Island. *Ibid.*, pp. 68-73 (Oct. 1).

Strauch, Friedrich. 1972. Phylogenese, Adaptation und Migration einiger nordischer mariner Molluskengenera (*Neptunea, Panomya, Cyrtodaria* und *Mya*). Abh. Senckenberg. Naturforsch. Ges., no. 531, pp. 1-211, 11 pls. (Nov. 15).

Strong, A. M. 1928. West American Mollusca of the genus *Phasianella*. Proc. Calif. Acad. Sci., ser. 4, vol. 17, no. 6, pp. 187-203, 1 pl. (June 22).

Swan, E. F., and J. H. Finucane. 1952. Observations on the genus *Schizothaerus*. Nautilus, vol. 66, no. 1, pp. 19-26, pls. 2-4 (July).

Taylor, D. W., and N. F. Sohl. 1962. An outline of gastropod classification. Malacologia, vol. 1, no. 1, pp. 7-32.

Tesch, J. J. 1949. Heteropoda: The Carlsberg Foundation's oceanographic expedition round the world, 1928-30, vol. 6 (Dana Rept. 34), 53 pp., 5 pls., 44 figs.

Thiele, Johannes. 1929-35. Handbuch der Systematischen Weichtierkunde. Jena: Fischer. Vol. 1, pt. 1, pp. 1-376, 470 figs. (1929); vol. 1, pt. 2, pp. 377-778, 313 figs. (1931); vol. 2, pt. 3, pp. 779-1034, 110 figs. (1934); vol. 2, pt. 4, pp. 1035-1154, 4 figs. (1935). (Reprinted 1963, Steickert-Hafner, New York.)

Thompson, T. E. 1971. Tritoniidae from the North American Pacific Coast (Mollusca: Opisthobranchia). Veliger, vol. 13, no. 4, pp. 333-38, 3 figs. (Apr. 1).

Tikasingh, E. S., and Ivan Pratt. 1961. The classification of endoparasitic gastropods. System. Zool., vol. 10, no. 2, pp. 65-69.

Turner, Ruth. 1954-55. The family Pholadidae in the western Atlantic and the eastern Pacific. Pt. I: Pholadinae. Johnsonia, vol. 3, no. 33, pp. 1-64, figs. 1-34 (May 17, 1954). Pt. II: Martesiinae, Jouannetiinae, and Xylophaginae. *Ibid.*, vol. 34, pp. 65-160, figs. 35-93 (Mar. 29, 1955).

——— 1966. A survey and illustrated catalogue of the Teredinidae. Mus. Comp. Zool. Harvard Univ., 265 pp., 64 pls.

Van der Spoel, S. 1967. Euthecosomata: A group with remarkable developmental stages (Gastropoda, Pteropoda). J. Noordunn en Zoon N.V., Gorinchem, Netherlands. 375 pp., 366 figs. (Nov. 20).

Vokes, H. E. 1967. Genera of the Bivalvia: A systematic and bibliographic catalogue. Bulls. Amer. Paleont., vol. 51, no. 232, 394 pp.

Voss, N. A. 1970. A monograph of the Cephalopoda of the North Atlantic: The family Histioteuthidae. Bull. Marine Sci., vol. 19, no. 4, pp. 713-867, 37 figs. (Jan. 12). Also includes Pacific forms.

Young, Richard E. 1972. The systematics and areal distribution of pelagic cephalopods from the seas off southern California. Smithson. Contrib. Zool., no. 97, 159 pp., 38 pls., 15 figs.

INDEX

INDEX

This index lists all taxa above species level to be found in the text. (The species given in the Identification of Figures, pp. 165-70, are not included here, but are arranged alphabetically by genus within that section.) For taxa below the family level, page references falling within the Systematic Lists (pp. 131-50) are followed in parentheses by the number of the family they are included in. Because of the many problems in synonymy that are still with us, the Index makes no attempt to differentiate between genera and subgenera, though both are given in italic type.